T0205574

Movement of Acoustic Energy in the Ocean

Vladimir A. Shchurov

Movement of Acoustic Energy in the Ocean

Vladimir A. Shchurov
Department of Ocean Acoustics
V. I. Il'ichev Pacific Oceanological Institute
Far Eastern Branch of Russian Academy
of Sciences
Vladivostok, Russia

ISBN 978-981-19-1302-0 ISBN 978-981-19-1300-6 (eBook)
https://doi.org/10.1007/978-981-19-1300-6

This Springer imprint is published by the registered company Springer Nature Singapore Pte Ltd.
The registered company address is: 152 Beach Road, #21-01/04 Gateway East, Singapore 189721, Singapore

Preface by Vladimir A. Shchurov: From Publication in Russian, 2019 (Translated from the Russian)

The monograph summarises many years of experimental vector-phase investigations of specific shallow waters (Sea of Japan shelf and the Kuril–Kamchatka Chain) and deep waters of the open ocean (northwestern and central Pacific and southern Indian Ocean). The Pacific Oceanological Institute first saw the need for vector-phase studies in late 1970s while studying the anisotropy of low-frequency underwater acoustic noise as part of the Metrology research project. Vector-phase studies were first conducted in northwestern Pacific in 1978–1979 aboard the R/V *Callisto*. That marked the beginning of development of vector-phase systems for fundamental and applied acoustic research of the ocean by the Ocean Acoustic Noise Laboratory of the Pacific Oceanological Institute, then part of the USSR Academy of Sciences. Research vessels *Callisto*, *Akademik Vinogradov*, *Akademik Lavrentyev* and hydrographic vessels of the Pacific Fleet served as the platform for the studies.

The expeditions developed, perfected and made use of the vector-phase equipment created by the Pacific Oceanological Institute's Ocean Acoustic Noise Laboratory. The author's early monographs *Vector Acoustics of the Ocean* (Russian edition, 2003), *Vector Acoustics of the Ocean* (English edition, 2006) and *Ocean Acoustics* (Mandarin edition, 2010) differ in content but refer to the same avenue of research in their titles. This monograph, *Movement of Acoustic Energy in the Ocean*, describes new results and summarises the ones described earlier. The data presented in the four monographs lay the groundwork for vector ocean acoustics as an independent branch of modern underwater physical acoustics. The purpose of this monograph is to use field experiment to demonstrate the fundamental significance of the phenomena discovered by the author and their effect on the transfer of signal energy in the ocean. For example, phenomena such as compensation of reciprocal energy fluxes and vortex structures have scarcely been studied, since their detection requires vector-phase analysis. Using these phenomena in applications is a matter for the future of modern vector acoustics.

The monograph consists of six chapters.

Chapter 1 outlines the mathematical tools necessary for understanding the physics of acoustic phenomena and data processing. The mathematical description is presented in deterministic form for the case of a homogeneous infinite space.

According to the virial theorem, these expressions also hold true in the real ocean environment for statistical averages. The four channels of the combined receiver $p(t)$, $V(x, y, z, t)\{V_x, V_y, V_z\}$ are processed jointly using FFT and Hilbert transforms.

Processing of random numerical series is mostly digital and multiplicative. This reduces digital processing to correlation analysis of data with a time shift between channels $\tau = 0$. We use the correlation theory of coherence developed in optics and radiophysics for statistical data processing and analysis of physical phenomena in the acoustic field of the real ocean.

The signal energy flux density vector in our analysis is expressed in terms of intensity components, i.e. through second-order statistics. Processing experimental data requires a study of space-time relations in the field of signal energy flux density, underwater ambient noise and interference noise. To do this, we need to transcend the correlation theory and move on to a fourth-order statistical moment—i.e. to investigate correlations among second-order statistics. This has pushed the boundaries of the statistical theory behind the processing of time- and space-dependent data. Building on the fourth-order statistical moment theory, we designed an experimental instrument equivalent to the Young–Rayleigh optical interferometer and used it in experiment. Drawing an analogy with stellar interferometry, this area may hold promise for underwater acoustics in fundamental as well as applied research and doubtless merits further study.

Chapter 2 outlines the basics of the theory of vector-phase measurements and describes equipment developed by the laboratory team. The hydrophone has been the default acoustic receiver in scientific research ever since the French scientist Langevin used it to detect submarines in World War I. However, to fully describe an acoustic field, we need to know these eight variables: three orthogonal components of the fluid particle velocity vector in the acoustic wave $V(x, y, z, t)$ $\{V_x, V_y, V_z\}$, two thermodynamic scalars: acoustic pressure $p(x, y, z, t)$, density of the medium $\rho(x, y, z, t)$ and three differential phase relationships $\Delta\varphi_{pVx}$, $\Delta\varphi_{pVy}$, $\Delta\varphi_{pVz}$. In the fundamental equations of acoustics, density of the medium ρ is usually set equal to the unperturbed density of fluid at rest ρ_0. Therefore, scalar characteristics are not enough, and the vector-phase method is a necessity in acoustic research. The Ocean Acoustic Noise Laboratory of the Pacific Oceanological Institute first applied the vector-phase methodology in a real ocean in 1979 to study low-frequency underwater acoustic noise. Initially (1978–1980), the philosophy and technique of the vector-phase method relied on research by the Acoustics Department of the Physics Faculty at Moscow State University. The collaborative first vector-phase studies of anisotropy of underwater low-frequency ambient noise were conducted in northwestern Pacific and near the Kuril–Kamchatka Chain in 1978–1980.

The vector-phase equipment and data processing algorithms that the laboratory developed in later years for fundamental and applied research have enabled investigations of open ocean and shallow sea in the 1–1000 Hz low-frequency region; this monograph presents their findings. To the author's knowledge, in the United States such studies used free-drifting neutrally buoyant floats in the frequency range of 0.4–20 Hz (the SWARM 95 experiment).

An important achievement of the study of energy movement in the ocean is the experimental discovery of compensation of reciprocal energy fluxes (Chap. 3). In this phenomenon, compensation is possible between coherent flows (tone or broadband signal) and underwater ambient noise—such as dynamic ocean noise or remote shipping noise, i.e. this phenomenon is unrelated to interference. In noise immunity terms, a combined receiver using this phenomenon can gain up to ~30 dB on a square-law detector using a hydrophone in anisotropic noise. Compensation of reciprocal energy fluxes was first tentatively noted in 1980 during processing of experimental data from the R/V *Callisto* expedition but was dismissed as a measurement error. A successful experiment in central Pacific shed light on the physics of this process. This phenomenon can be used for weak signal detection regardless of interference noise pattern. The work program of the passive vector-phase sonar employs this phenomenon (see Chap. 5).

The author discovered vortices of the acoustic intensity vector in the far field of a source in 2008 in the Peter the Great Gulf of the Sea of Japan (Chap. 4). The vortices had been predicted theoretically in 1989 (Iu. A. Kravtsov et al.). Between 1989 and 2008, vortices of the acoustic intensity vector were not discussed in scientific literature because no one observed them in experiments during that time. Our publications on the subject in 2010 gave rise to first works by other authors published from 2013. A salient feature of the vortices is rotation of the signal energy flux vector towards its source. The vortices are generated by modal interference, meaning that the field's modal structure can be studied by observing the movement of the vortices. As is known, the acoustic field is a potential field and the particle velocity vector is vortex-free, i.e. $\mathbf{rot}\ V = 0$. Vortex structures of the energy flux density vector arise in complex interference fields of the shallow sea, with $\mathbf{rot}\ pV \neq 0$. We have established that a near-deterministic periodic pattern of vertical energy flows arises when vortices and vorticity form in the vertical plane, which contradicts the mode theory. Stationary vortices and vortices oscillating along the horizontal axis of the waveguide were discovered. This phenomenon appears to be associated with intermodal interference among different number modes when the hydrophysical situation in the waveguide changes. This effect requires further study.

The combined receiver's noise immunity advantage with multiplicative signal processing over the square-law detector based on a hydrophone (Chap. 5) sparked a lengthy debate between M. D. Smaryshev and V. A. Shchurov in the *Akusticheskij Zhurnal* (*Acoustical Physics*) in 2005. In the end, our estimate of noise immunity prevailed: SNR(PV) ~16–20 dB in a diffuse field and ~30 dB in an anisotropic field, which we published in a 2002 paper. Given their high noise immunity compared with square-law detectors, combined receivers can be used in next-generation underwater detection systems. The compensation effect can simplify statistical detection, since the signal arrival criterion is known in advance. A feature of the combined receiver, the 180° phase jump when the acoustic source crosses over the minimum of the dipole pattern, further simplifies the statistical detection problem. The vector-phase passive sonar described in the monograph can be a component in a state-of-the-art detection system.

A word on terminology. The author stands by the term 'vector-phase method' coined at the Acoustics Department of the Faculty of Physics at Moscow State University (S. N. Rzhevkin, L. N. Zakharov, V. I. Sizov, 1961). Importantly, Russian terminology of the vector-phase method has been fully adopted in the United States and elsewhere (W. A. Kuperman, G. L. D'Spain, USA). For example, the vector-phase method used to be called the intensity measurement method in the United States, and the vector receiver a particle velocity hydrophone. That said, terms such as 'scalar–vector method' or 'vector–scalar method' are to be found in current Russian scientific literature, which can be attributed to the authors not entirely appreciating the depth of the physics of vector processes in the acoustic field of the ocean. This can, in part, be explained by the fact that this nomenclature is preferred by acousticians whose background is in engineering rather than in physics. The term 'vector acoustics of the ocean' is a logical extension of the term 'vector-phase method', which is used to study the vector parameters of the acoustic field. According to the mathematical definition of scalar and vector fields (I. N. Bronshtein, K. A. Semendyayev, *Handbook of Mathematics* (in Russian), 1981), the author considers it logical to designate this area of underwater physical acoustics 'vector acoustics of the ocean'.

The Pacific Oceanological Institute can claim much of the credit for developing the vector-phase method and for its achievements in Russia. Chapter 6 presents the equipment (mock-ups) developed at the Pacific Oceanological Institute and its use in scientific experiments from 1978 onwards, along with some details of ocean expeditions and international conferences.

The author wishes to express his sincere gratitude to Fellows of the Russian Academy of Sciences V. A. Akulichev and G. I. Dolgikh, to V. B. Lobanov for fruitful discussions and support, and to researchers at the Ocean Acoustic Noise Laboratory E. S. Tkachenko, A. S. Liashkov, Iu. A. Khvorostov and S. G. Shcheglov, whose professionalism has made complex acoustic experiments in the ocean possible.

Vladivostok, Russia Vladimir A. Shchurov

Acknowledgements

Since the publication of the first edition in 2019, in Russian, the monograph by Vladimir A Shchurov titled *Movement of Acoustic Energy in the Ocean* has enhanced the reputation of the author as being the foremost international researcher on the application of vector acoustics for underwater acoustics studies. Australian Acoustical Society (AAS) Publications is an expansion of the current successful journal Acoustics Australia. AAS Publications is honoured to have had the opportunity to work with Vladimir Shchurov and to publish in English, via Springer, this important contribution to the understanding of how sound propagates in the ocean.

This English translation has been undertaken by Mykhaylo Syzonenko from Australian Multilingual Translation Services Pty Ltd. The liaison provided by Prof. Alexander Gavrilov from the Centre for Marine Science and Technology at Curtin University, Western Australia with both the author and translator has been instrumental in producing this English translation, as well as the technical and organisational support from the Australian Acoustical Society with funding provided by the US Office of Naval Research. Editor-in-Chief for Acoustics Australia, Associate Professor Marion Burgess from the University of New South Wales has been responsible for working with all involved to bring this translation to publication.

From the Editor of Publication in Russian, 2019 (Translated from the Russian)

Movement of Acoustic Energy in the Ocean continues a series of three monographs by the author entitled *Vector Acoustics of the Ocean*. Published in Russian (2003), English (2006) and Mandarin (2010), the previous monographs were devoted to experimental vector-phase studies of the shallow sea and deep open ocean. These are pioneering works in this field. This monograph, *Movement of Acoustic Energy in the Ocean*, is being published under the caption of 'Vector acoustics of the ocean' to emphasise the unity of the research method, which has put vector acoustics of the ocean on the map as a branch of underwater acoustics in its own right. This nomenclature is also used by many authors outside Russia.

What makes the monograph pertinent is that it casts acoustical processes in terms of data that are necessary to completely describe phenomena in the real ocean. Numerous phenomena were discovered as a result that had not been observed in the ocean environment. The author's original work on vortex structures broke new ground. The vortex structures discovered by the author in 2008 have drastically changed our perspective on signal energy movement in the shallow sea waveguide. The discovered vertical flows of signal energy (which contradict the mode theory), horizontal oscillatory motion of vortices, the phenomenon of compensation of reciprocal energy fluxes and the vector-phase sonar—all this the author seamlessly weaves into the big picture of energy movement in the ocean environment. The fundamental phenomena of the acoustic vector field presented in the book and the case for their use in applications are without doubt what makes the monograph valuable.

Movement of Acoustic Energy in the Ocean reflects the advances made by underwater physical acoustics and will be useful to researchers in the field.

September 2019

G. I. Dolgikh
Fellow of the Russian Academy of Sciences
Vladivostok, Russia

Contents

Chapter 1
Vector Representation of the Acoustic Field

1.1 Introduction

To completely describe the movement of acoustic energy through a continuous medium, we need to know the following physical quantities: medium density ρ, acoustic pressure $p(t)$, three orthogonal components of the acoustic particle velocity $V(t)\{V_x(t), V_y(t), V_z(t)\}$ and differential phase relations between these quantities and the orthogonal components of the particle velocity. All these values figure in linearised equations of fluid dynamics. Equations of motion of energy in continuous media were first formulated by Nikolai Umov (1846–1915), a foremost Russian scientist. In 1873, he introduced the energy flux density vector, a foundational concept of modern-day physics, which was later called the Umov vector. The intensity vector is the averaged Umov vector. Vector acoustics studies these physical quantities at the same point in time and space of the ocean environment.

We will retronymically call 'scalar acoustics' that part of acoustic that deals with the scalar acoustic pressure $p(t)$.

In this chapter, we will present physical relationships that describe the acoustic field in an infinite uniform space using deterministic harmonic functions. The deterministic approach is justified when describing tonal (monochromatic) signals in complex acoustics fields of the real ocean. The experimentally measured values of acoustic pressure $p(x, y, z, t)$ and orthogonal components of the medium particle speed vector $\boldsymbol{V}(t)\{V_x(t), V_y(t), V_z(t)\}$ usually take the form of random time series. Statistical analysis leads us to average characteristics of the acoustic field, which can be accurately described by deterministic functions [1–4].

V. A. Shchurov, *Movement of Acoustic Energy in the Ocean*,
https://doi.org/10.1007/978-981-19-1300-6_1

1.2 Scalar and Vector Characteristics of the Acoustic Field

For a flat wave travelling through a homogeneous infinite space in the adiabatic approximation, we will express the energy density and energy flux density according to [3, 4].

Instantaneous energy density of the acoustic field $E(t)$ is the sum of instantaneous kinetic $E_k(t)$and instantaneous potential $E_p(t)$ energies:

$$E(t) = E_k(t) + E_p(t) = \frac{1}{2}\rho V^2(t) + \frac{1}{2}\frac{p^2(t)}{\rho c^2}. \tag{1.1}$$

In a flat wave travelling through a homogeneous infinite space, $p(t) = \pm\rho c V(t)$ at any time and at any point in the wave. In this case, the total energy density $E(t)$ of a flat wave can be written as

$$E(t) = \rho V^2(t) = \frac{p^2(t)}{\rho c^2}. \tag{1.2}$$

It follows from (1.2) that in a flat travelling wave, the density of kinetic energy equals the density of potential energy at any point and at any time.

For an arbitrary wave, the equivalent of (1.2) can only be written for time-averaged total energy. This follows from a general theorem of mechanics that states that in any system experiencing small oscillations, the average potential energy equals the average kinetic energy [4].

The vector of instantaneous intensity of a plane wave (vector of instantaneous energy flux density—the Umov vector) has the form [6]:

$$\boldsymbol{j} = p(t)V(t)\boldsymbol{n}. \tag{1.3}$$

In (1.1)–(1.3), $p(t)$ and $V(t)$ are instantaneous acoustic pressure and particle velocity vector, respectively; ρ is the density of the medium; c is the speed of sound; and $\boldsymbol{n}$ is a unit vector in the wave direction. In a flat monochromatic wave travelling in the +x direction, acoustic pressure and particle velocity have the same phase:

$$\begin{aligned} p(t) &= p_0 \cos(\omega t - kx - \varphi_p), \\ V(t) &= V_0 \cos(\omega t - kx - \varphi_p), \end{aligned} \tag{1.4}$$

where p_0 and V_0 are amplitudes of pressure and particle velocity, respectively; ω is the angular frequency; t is time; k is the wavenumber; and φ_p is the initial phase of $p(t)$ and $V(t)$. The highest pressure corresponds to the highest particle velocity in the $+x$ direction, the lowest pressure to the highest velocity in the $-x$ direction.

The instantaneous energy density according to (1.1), (1.4) is

$$E(t) = \frac{1}{2}\rho V_0^2 \cos^2(\omega t - kx - \varphi_p) + \frac{1}{2}\frac{p_0^2}{\rho c^2}\cos^2(\omega t - kx - \varphi_p). \tag{1.5}$$

Both time- and space-averaged energy density in a travelling monochromatic flat wave equal

$$E = \frac{1}{2}\rho V_0^2 = \frac{1}{2}\frac{p_0^2}{\rho c^2}. \tag{1.6}$$

Instantaneous intensity of a monochromatic flat travelling wave

$$I(t) = p(t)V(t) = \frac{1}{2}p_\theta V_0 + \frac{1}{2}p_0 V_0 \cos 2(\omega t - kx - \varphi_p). \tag{1.7}$$

The first term in (1.7) is independent of the time t. The second term vanishes over a time equal to one period or a multiple of the period. Therefore, the average intensity (or simply intensity) of the flat wave becomes

$$\boldsymbol{I} =< I(t)\boldsymbol{n} >= \frac{1}{2}p_0 V_0 \boldsymbol{n}. \tag{1.8}$$

Intensity is the amount of sound energy carried per unit time of 1 s across a unit area of 1 m^2 of the wave surface in the wave direction. This means that intensity is a vector, with a magnitude and a direction. The unit of intensity is J/s m^2 = W/m^2. The unit of energy density is J/m^3.

Since $p(t) = \pm\rho c V(t)$, the intensity of a single flat travelling monochromatic wave can be recast as follows:

$$\boldsymbol{I} = \frac{1}{2}p_0 V_0 \boldsymbol{n} = \frac{1}{2}\rho c V_0^2 \boldsymbol{n} = \frac{1}{2}\frac{p_0^2}{\rho c}\boldsymbol{n}. \tag{1.9}$$

From (1.6) and (1.9) it follows that $I = Ec$, i.e. in a plane wave energy flux density equals energy density times the sound speed.

If several flat waves of the same frequency arrive at the observation point from different directions, then the net particle velocity in the general case is shifted in phase relative to the phase of acoustic pressure, and the net particle velocity vector does not coincide with the wave direction [3].

In this case, we will write the oscillations of the four components of the acoustic field $p(t)$, $V_x(t)$, $V_y(t)$, $V_z(t)$ in the form:

$$\begin{aligned} p(t) &= p_0(\omega t + \varphi_p), \\ V_x(t) &= V_{0,x}\cos(\omega t + \varphi_p - \varphi_x), \\ V_y(t) &= V_{0,y}\cos(\omega t + \varphi_p - \varphi_y), \\ V_z(t) &= V_{0,z}\cos(\omega t + \varphi_p - \varphi_z), \end{aligned} \tag{1.10}$$

where p_0, $V_{0,x}$, $V_{0,y}$, $V_{0,z}$ are amplitudes, ω is the angular frequency; t is the time; φ_p is the initial phase of acoustic pressure; $(\varphi_\mathrm{p} - \varphi_x)$, $(\varphi_\mathrm{p} - \varphi_y)$, $(\varphi_\mathrm{p} - \varphi_z)$ are phase differences between acoustic pressure and the x, y, z components of particle velocity.

The net particle velocity vector for (1.10) has the form:

$$\boldsymbol{V}(t) = \boldsymbol{i}V_x(t) + \boldsymbol{j}V_y(t) + \boldsymbol{k}V_z(t), \tag{1.11}$$

where $\boldsymbol{i}, \boldsymbol{j}, \boldsymbol{k}$ are Cartesian unit vectors.

In the case of superposition of plane waves, amplitudes p_0, $V_{0,x}$, $V_{0,y}$, $V_{0,z}$ and phase differences $(\varphi_\mathrm{p}-\varphi_x)$, $(\varphi_\mathrm{p}-\varphi_y)$, $(\varphi_\mathrm{p}-\varphi_z)$ are determined by the interference pattern of the field and are functions of coordinates.

In stationary fields, these quantities are time-independent. Particle velocity vector $\boldsymbol{V}$ at a given point of the field can be represented as the sum of two vectors [3]:

$\boldsymbol{V}_{\mathrm{ac}}$ is the active component of the particle velocity;
$\boldsymbol{V}_{\mathrm{reac}}$ is the reactive component of the particle velocity:

$$\begin{aligned} \boldsymbol{V}_{\mathrm{ac}} &= \boldsymbol{i}V_{0,x}(t)\cos\varphi_x + \boldsymbol{j}V_{0,y}(t)\cos\varphi_y + \boldsymbol{k}V_{0,z}(t)\cos\varphi_z, \\ \boldsymbol{V}_{\mathrm{reac}} &= \boldsymbol{i}V_{0,x}(t)\sin\varphi_x + \boldsymbol{j}V_{0,y}(t)\sin\varphi_y + \boldsymbol{k}V_{0,z}(t)\sin\varphi_z, \end{aligned} \tag{1.12}$$

Relations (1.10) describe an ellipse, which is the trajectory of the vector $\boldsymbol{V}(t)$ at a given point of the field and in the plane determined by time-independent vectors $\boldsymbol{V}_{\mathrm{ac}}$ and $\boldsymbol{V}_{\mathrm{reac}}$ [3, 5]. Therefore, in an acoustic field that is a superposition of flat deterministic monochromatic waves, the relationships between $\boldsymbol{V}_{\mathrm{ac}}$ and $\boldsymbol{V}_{\mathrm{reac}}$ are determined by the phase difference between pressure and particle velocity.

Both active and reactive components of the acoustic field contribute to its total energy density. However, intensity of the acoustic field depends only on the active component, while the reactive component has zero intensity. One example of a deterministic reactive field is a standing wave. A diffuse field, whose energy flux density (intensity) is also zero, cannot be viewed as a reactive field because it is the product of averaging random processes in acoustic fields, while active and reactive fields are a property of deterministic monochromatic fields only.

The averages of components I_x, I_y and I_z of intensity of a sum of monochromatic waves in Cartesian coordinates x, y, z can be written as:

$$\begin{aligned} I_x &= \frac{1}{2}p_0 V_{0,x}\cos(\varphi_p - \varphi_x), \\ I_y &= \frac{1}{2}p_0 V_{0,y}\cos(\varphi_p - \varphi_y), \\ I_z &= \frac{1}{2}p_0 V_{0,z}\cos(\varphi_p - \varphi_z), \end{aligned} \tag{1.13}$$

where p_0, $V_{0,x}$, $V_{0,y}$, $V_{0,z}$ are amplitudes of the sum of monochromatic waves of acoustic pressure and orthogonal components *x, y, z* of particle velocity at the observation point; $(\varphi_p - \varphi_x)$, $(\varphi_p - \varphi_y)$, $(\varphi_p - \varphi_z)$ are phase differences between the acoustic pressure and the orthogonal components of particle velocity.

The net average vector of energy flux density of the sum of monochromatic waves of the same frequency can be written as:

$$\boldsymbol{I} = \boldsymbol{i} I_x + \boldsymbol{j} I_y + \boldsymbol{k} I_z, \tag{1.14}$$

where $\boldsymbol{i}, \boldsymbol{j}, \boldsymbol{k}$ are unit vectors along the *x, y* and *z axes*, respectively.

It follows from (1.13) that when a phase difference equals $\pi/2$, the energy flux density (1.14) or any of its components vanish. In the case of superposition of deterministic monochromatic waves, this condition is met for the region of space where standing waves form. In a standing wave, energy density of the acoustic field is concentrated in its reactive component, but intensity is zero.

In the complex representation of plane monochromatic waves, the average intensity (averaged over the period or a multiple of the period) can be written in the form:

$$I = \frac{1}{2}\langle \mathrm{Re}\ p(t)V^*(t)\rangle = \frac{1}{2}\langle \mathrm{Re}\ p^*(t)V(t)\rangle, \tag{1.15}$$

where Re denotes the real part of a complex quantity and * means complex conjugation.

Expressions $p(t)\ V^*(t)$ and $p^*(t)\ V(t)$ differ, in that the former refers to negative and the latter to positive frequencies.

The real parts of these complex expressions are equal. It then follows that waves differing in the frequency sign are identical physical entities [3, 5].

In the future, we will use the expression

$$I = \frac{1}{2}\langle \mathrm{Re}\ p(t)V^*(t)\rangle, \tag{1.16}$$

Orthogonal components of the energy flux density vector $\boldsymbol{I}\{I_x, I_y, I_z\}$ at an observation point $A(x,y,z)$ [(1.13)] in complex notation will have the form:

$$\begin{aligned} I_x &= \frac{1}{2}\langle \mathrm{Re}\ p(t)V_x^*(t)\rangle, \\ I_y &= \frac{1}{2}\langle \mathrm{Re}\ p(t)V_y^*(t)\rangle, \\ I_z &= \frac{1}{2}\langle \mathrm{Re}\ p(t)V_z^*(t)\rangle, \end{aligned} \tag{1.17}$$

All relations given in this section hold true for deterministic monochromatic signals and a plane wave in a homogeneous infinite space.

1.3 Differential Phase Relationships in Complex Acoustic Vector Fields

Consider a plane wave travelling in the + x direction. Acoustic pressure $p(t)$ and component $V_x(t)$ of the particle velocity can be expressed as:

$$p(t) = A\cos(\omega t - kx + \varphi), \tag{1.18}$$

$$V_x(t) = (A/\rho c)\cos(\omega t - kx + \varphi) = p/\rho c. \tag{1.19}$$

Thus in the case of infinite homogeneous space in an acoustic wave travelling in the + x direction, acoustic pressure $p(t)$ and particle velocity $V_x(t)$ have the same phase (the phase difference is zero) and are related by the equality $p(t) = \rho c V(t)$. In a wave travelling in the – x direction, acoustic pressure $p(t)$ and particle velocity $V_x(t)$ have opposite phases (are 180° out of phase), i.e. $p(t) = -\rho c V(t)$ (where ρ is the density of the medium and c is the sound speed).

It follows from the above that phase differences $\Delta\varphi_x = \varphi_p - \varphi_x$, $\Delta\varphi_y = \varphi_p - \varphi_y$, $\Delta\varphi_z = \varphi_p - \varphi_z$ between pressure $p(t)$ and the orthogonal components of particle velocity will be zero if the wave travels in the $+x$, $+y$ or $+z$ direction and will equal 180° if the wave travels in the opposite direction. Hence, for example, it follows that when a sound source moving in the $x0y$ plane from quadrant II to quadrant I crosses the y axis, the phase difference $\Delta\varphi_x$ between acoustic pressure $p(t)$ and particle velocity component $V_x(t)$ must jump from 0 to 180°.

A combined receiver can measure acoustic pressure $p(t)$ and the three orthogonal components of particle velocity $V_x(t)$, $V_y(t)$, $V_z(t)$ or acceleration $a_x(t)$, $a_y(t)$, $a_z(t)$ jointly and simultaneously [7].

Usually, the x and y axes of a combined receiver are horizontal, and the z axis is vertical and points from the surface to the bottom. A wave's direction is specified by angles ψ and θ. Azimuthal angle ψ is measured in the $x0y$ plane from the + x direction. Polar angle θ is measured from the + z direction. Depending on ψ and θ, the phase differences are:

$$\Delta\varphi_x = \begin{cases} 0^\circ, \; 270^\circ < \psi < 90^\circ; \\ \pi, \; 90^\circ < \psi < 270^\circ; \end{cases}$$
$$\Delta\varphi_y = \begin{cases} 0, \; 0^\circ < \psi < 180^\circ; \\ \pi, \; 180^\circ < \psi < 360^\circ; \end{cases}$$
$$\Delta\varphi_z = \begin{cases} 0^\circ, \; 0^\circ < \theta < 90^\circ; \\ \pi, \; 90^\circ < \theta < 180^\circ; \end{cases} \tag{1.20}$$

Phase differences (1.20) uniquely determine the octant of the sound source.

The $a_x(t)$ component of particle acceleration in a plane wave is determined by:

$$a_x(t) = \frac{\mathrm{d}V_x}{\mathrm{d}t} = \frac{A\omega}{\rho c}\cos\left(\omega t - kx + \varphi + 90^\circ\right). \tag{1.21}$$

This means that particle acceleration $a_x(t)$ has a phase shift of 90° relative to $p(t)$ and $V_x(t)$. In this sense, when the wave direction moving in the $x0y$ plane crosses the y axis from the first quadrant to the second, the phase difference between $p(t)$ and $a_x(t)$ will jump from + 90 to – 90°, i.e. the phase difference will also jump 180°.

As full-scale experiment shows, the phase difference jump is observed for both tonal and noise signals, as well as for pulses of arbitrary shape [7]. Let us look at the jumps in phase difference experimentally measured for a source moving around a combined receiving system. The source of noise signal in the experiment was a motorboat circling at a constant speed. The z axis of the receiving system was a vertical line passing through the centre of the circle. The combined receiving system was located at a depth of 30 m.

Figure 1.1 shows the results of the experiment for a noise signal in the $\Delta f =$ 600–800 Hz band.

When $\psi(t) = 0°$, the wave travels in the $+ x$ direction; when $\psi(t) = 90°$, in the + y direction; when $\psi(t) = 180°$, in the $- x$ direction (Fig. 1.1a), etc. Comparing Fig. 1.1a–c we see that phase differences $\Delta\varphi_x(t)$ and $\Delta\varphi_y(t)$ experience jumps between + 90° and – 90° when the wave propagation vector crosses over from one quadrant to another. For a plane harmonic wave from a single source propagating through a homogeneous infinite space, the average flux density vector is expressed by

$$\begin{aligned} \boldsymbol{I} &= \langle p(t)\boldsymbol{V}(t)\rangle = \frac{1}{2}pV\boldsymbol{n} \\ &= \frac{1}{2}\rho c V^2\boldsymbol{n} = \frac{1}{2}\frac{p^2\boldsymbol{n}}{\rho c}, \end{aligned} \tag{1.22}$$

where p and V are the amplitudes of pressure and particle velocity and $\boldsymbol{n}$ is the unit vector in the wave direction, $|\boldsymbol{n}| = 1$.

If the acoustic field at some point in space is a superposition of k plane statistically independent harmonic waves of the same frequency arriving from different directions $\boldsymbol{n}_i$, then the net averaged energy flux along a direction $\boldsymbol{r}$ is the sum of k projections of individual fluxes onto the direction $\boldsymbol{r}$:

$$I_r = \frac{1}{2}\sum_{i=1}^{k} p_i V_i \cos\theta_i, \tag{1.23}$$

where p_i and V_i are pressure and particle velocity amplitudes of the ith wave, and θ_i is the geometric angle between the particle velocity vector of the ith plane wave and the direction $\boldsymbol{r}$ [8].

Let us recast (1.23) in terms of net pressure $p(t)$ and net particle velocity $V_r(t)$ along a direction $\boldsymbol{r}$ in the simplest case of superposition of two intersecting waves. We will express acoustic pressures $p_1(t)$ and $p_2(t)$ of these wes and their corresponding particle velocities $V_1(t)$ and $V_2(t)$ as follows:

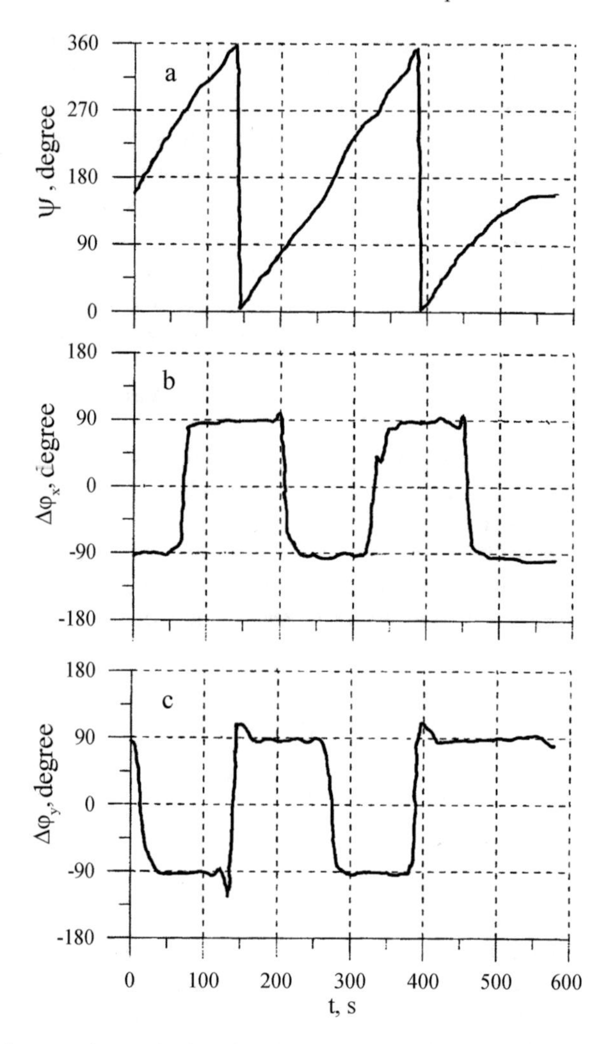

Fig. 1.1 Synchronous changes in the azimuthal angle $\psi(t)$ and $\Delta\varphi x\ (t)$, $\Delta\varphi y(t)$ in an experiment with the sound source moving around a combined receiving system in the horizontal plane: pane **a** azimuthal angle $\psi(t)$ between $+\ x$ and the acoustic wave direction; pane **b** phase difference $\Delta\varphi x(t) = \varphi p - \varphi x$; pane **c** phase difference $\Delta\varphi y(t) = \varphi p - \varphi y$. Phase differences $\Delta\varphi x(t)$ and $\Delta\varphi y(t)$ are between pressure $p(t)$ and components of particle acceleration $ax(t)$ and $ay(t)$

$$
\begin{aligned}
p_1(t) &= p_1 \cos(\omega t + \psi_1), \\
p_2(t) &= p_2 \cos(\omega t + \psi_2), \\
V_1(t) &= V_1 \cos(\omega t + \psi_1), \\
V_2(t) &= V_2 \cos(\omega t + \psi_2),
\end{aligned}
\tag{1.24}
$$

where p_1, p_2, V_1, V_2 are amplitudes and ψ_1 and ψ_2 are initial phases of the oscillations.

The net pressure $p(t)$ is the algebraic sum of $p_1(t)$ and $p_2(t)$:

$$p(t) = p\cos(\omega t + \varphi_{\mathrm{p}}), \tag{1.25}$$

where p is the pressure amplitude of the net acoustic wave and φp is the phase angle of the net acoustic pressure.

The net particle velocity along the direction $\boldsymbol{r}$ will depend on the angles θ_1 and θ_2 between the directions of the incident waves and the direction $\boldsymbol{r}$:

$$V_r(t) = V_r\cos(\omega t + \varphi_{\mathrm{v}}) \tag{1.26}$$

where Vr is the amplitude of net particle velocity along the direction $\boldsymbol{r}$ and φ_{v} is the phase angle of net particle velocity.

Phase angles φ_{p} and φ_{v} equal:

$$\begin{aligned}\tan\varphi_{\mathrm{p}} &= \frac{p_1\sin\psi_1 + p_2\sin\psi_2}{p_1\cos\psi_1 + p_2\cos\psi_2},\\ \tan\varphi_{\mathrm{V}} &= \frac{(V_1\cos\theta_1)\sin\psi_1 + (V_2\cos\theta_2)\sin\psi_2}{(V_1\cos\theta_1)\cos\psi_1 + (V_2\cos\theta_2)\cos\psi_2},\end{aligned} \tag{1.27}$$

Component Ir expressed through the amplitudes of net acoustic pressure p and the component of the particle velocity Vr in the direction $\boldsymbol{r}$ can be written in the form:

$$I_r = \frac{1}{2}\left[pV_r\cos(\varphi_{\mathrm{p}} - \varphi_{\mathrm{V}})\right], \tag{1.28}$$

where $\varphi_{\mathrm{p}} - \varphi_{\mathrm{V}}$ is the phase difference between pressure and particle velocity in the resulting wave. Formulas (1.22), (1.23) and (1.28) express averages of the acoustic energy flux density vector over a period of time $t \geq T$, where T is the period of the harmonic wave. Therefore, if several flat waves travelling from different directions arrive at a point in space, then the net acoustic pressure $p(t)$ and the orthogonal components of net particle velocity $V_x(t)$, $V_y(t)$, $V_z(t)$ incur phase differences $\Delta\varphi_x = \varphi_{\mathrm{p}} - \varphi_x$, $\Delta\varphi_y = \varphi_{\mathrm{p}} - \varphi_y$, $\Delta\varphi_z = \varphi_{\mathrm{p}} - \varphi_z$, which may depend on time t.

Energy flux density vector $\boldsymbol{I} = \boldsymbol{i}I_x + \boldsymbol{j}I_y + \boldsymbol{k}I_z$ has its components I_x, I_y, I_z, which equal:

$$\begin{aligned}I_x &= \frac{1}{2}pV_x\cos(\varphi_{\mathrm{p}} - \varphi_x),\\ I_y &= \frac{1}{2}pV_y\cos(\varphi_{\mathrm{p}} - \varphi_y),\\ I_z &= \frac{1}{2}pV_z\cos(\varphi_{\mathrm{p}} - \varphi_z),\end{aligned} \tag{1.29}$$

where $p(t)$ and Vx, Vy, Vz are amplitudes of net acoustic pressure and of components of net particle velocity.

Energy flux at an observation point is an integral characteristic and is the sum of energy fluxes at a given frequency f_0 from k different acoustic sources. In this monograph, net energy flux is the sum of energy fluxes of flat travelling waves. This approach is justified if the receiver is in the wave zone (far field) of acoustic sources. The simplest sources of sound are a pulsating (monopole) and an oscillating (dipole) sphere.

If the signal is harmonic, phase velocity of spherical pressure waves (with a monopole as the acoustic source) is the same as phase velocity of flat waves.

As a result, expression $I = \frac{1}{2}\frac{p^2}{\rho c}$ holds true for spherical as well as for plane waves (1.22).

In a spherical wave, the relationship between pressure and particle velocity is more complex than in a flat wave. Particle velocity lags behind pressure by the angle φ, which is a function of distance to the source. In the far field, $p = \rho c V$ for a spherical wave (as well as for a plane wave), but pressure and particle velocity decrease inversely with increasing distance to the source. The expression for sound intensity of an emitting monopole in the far field is identical to that of the plane wave (1.22). For the acoustic field of a dipole, (1.22) holds only in the far field.

Since particle velocity in an acoustic wave is a vector, when two waves of the same frequency are superimposed, the net particle velocity vector must rotate. Knowing what plane the particle velocity vector moves in and what trajectory its end traces, we can uniquely determine the spatial vector of particle displacement, and therefore the path and direction of particle movement. In spatially complex acoustic fields, amplitudes and phases of particle velocity are functions of coordinates and are determined by the interference pattern in the field created by distributed sources, presence of boundaries, etc. In stationary acoustic fields, these are time-independent and can be derived. Thus by analogy with vector fields, such as electromagnetic field, we can apply the concept of 'polarisation' to acoustic fields in fluids to describe the behaviour of a sound wave at a point of the field [9].

The type of polarisation (linear, circular, elliptical) depends, among other things, on the angle θ between the instantaneous vectors of particle velocity $\boldsymbol{V}_1(t)$ and $\boldsymbol{V}_2(t)$ of two intersecting waves at a point in space (1.24). Superposition of two waves changes the direction of energy transfer in the acoustic field relative to the direction of net particle velocity. The angle θ is related to angles θ_1 and θ_2 [in (1.23)] by the expression $\theta = \theta_2 - \theta_1$. Therefore, differential phase relations $\Delta\varphi_x(t)$, $\Delta\varphi_y(t)$ and $\Delta\varphi_z(t)$ determine the type of polarisation of the net particle velocity vector in a complex acoustic field.

Generally, a real acoustic field is a function of coordinates and time (positions and distances to sources change, as do the magnitude and anisotropy of the noise field, etc.).

Acoustic fields of complex structure can be investigated more efficiently by measuring $\Delta\varphi_x(t)$, $\Delta\varphi_y(t)$ and $\Delta\varphi_z(t)$.

1.4 Instantaneous and Average Acoustic Intensity

By 'vector of instantaneous energy flux density' (Umov vector) and 'instantaneous intensity vector', we mean the following. In the theory of oscillations and waves, the temporal scale is the period $T = 2\pi/\omega$ and the spatial scale is the wavelength $\lambda = cT = 2\pi c/\omega$. Which is why by instantaneous values we mean the wavefield behaviour within a time interval of one period T and within a space interval of one wavelength. Therefore, the Umov vector shows whence and whereto field energy flows over a time of one period T on a segment of space equal to λ.

Let us derive the Umov vector for a harmonic (monochromatic) acoustic field. To do this, we will determine the instantaneous intensity of an acoustic field that is the sum of flat monochromatic waves. In the general case, a random number of plane waves arrive at a point in the acoustic waveguide, each taking its own path from the source to the receiver. Considering this field stationary and monochromatic of one angular frequency $\omega = 2\pi/T$, we can write the net oscillations of acoustic pressure and particle velocity at a point as follows:

$$\begin{aligned} p(t) &= p_0 \cos(\omega t + \varphi_{\mathrm{p}}), \\ \boldsymbol{V}(t) &= \boldsymbol{V}_0 \cos(\omega t + \varphi_{\mathrm{V}}), \end{aligned} \tag{1.30}$$

where p_0 is the net amplitude of the sound pressure; V_0 the net amplitude of the particle velocity; ϕp the phase of acoustic pressure; and φV the phase of particle velocity. In a flat sound wave, $p = \rho c V$.

Let us calculate the component of instantaneous intensity in a direction $\boldsymbol{d}$. In this case, the intensity is the product of instantaneous values of $p(t)$ and $V_{\mathrm{d}}(t)$, where $V_{\mathrm{d}}(t)$ is the component of particle velocity in the direction $\boldsymbol{d}$:

$$\begin{aligned} I_{\mathrm{d}}(t) &= p_0 V_{0,\mathrm{d}} \cos(\omega t + \varphi_{\mathrm{p}}) \cos(\omega t + \varphi_{\mathrm{V}}) \\ &= \frac{1}{2}\left[p_0 V_{0,\mathrm{d}} \cos(\varphi_{\mathrm{p}} - \varphi_{\mathrm{V}}) + p_0 V_{0,\mathrm{d}} \cos(2\omega t + \varphi_{\mathrm{p}} + \varphi_{\mathrm{V}})\right] \end{aligned} \tag{1.31}$$

The first term of the sum (1.31) is time-independent. At given p_0 and $V_{0,\mathrm{d}}$, it can vary depending on the magnitude and sign of the 'power factor' $-1 \leq \cos(\varphi_{\mathrm{p}} - \varphi_{\mathrm{V}}) \leq +1$. The second term shows that in the first quarter of the period $T = \omega/2\pi$ during which one fluctuation energy flows away from the source, and in the second quarter of the period it reverses direction and flows towards the source, and this process of energy transfer is repeated throughout the stationary field. To summarise, the first term of the sum (1.31) shows that throughout the period T energy flows from the source to the receiver, and the flux depends only on the 'dynamic' between $p(t)$ and $V_{\mathrm{d}}(t)$, i.e. their phase difference. The second term shows that the direction of energy flow reverses during a period. The acoustic field in a homogeneous infinite medium divides the entire space between the source and the receiver into segments one wavelength λ long. Within each segment, the acoustic energy will 'swing' back and forth over each period T, i.e. the second term of the sum is associated with the local spatial property of the acoustic field, and its magnitude will also depend on the field's phase characteristics. Averaging (1.31) over a multiple of the period loses the time dependence of intensity, and with it much of the information about the acoustic field:

$$I_{\mathrm{d}} = \frac{1}{t_0}\int_0^{t_0} p(t)V_{\mathrm{d}}(t)\mathrm{d}t = p_0 V_{0,\mathrm{d}} \cos(\varphi_{\mathrm{p}} - \varphi_{\mathrm{V}}). \tag{1.32}$$

Expression (1.32) represents the part of the energy that the acoustic field carries at a point in the direction $\boldsymbol{d}$ over the time t_0, and equals the average intensity (or simply intensity). Naturally, at some interval of time $t \gg t_0$ or when the position of the observation point changes, acoustic field intensity (1.32) will depend on $\boldsymbol{r}$ and t, i.e. $I_{\mathrm{d}}(\boldsymbol{r}, t)$.

In Cartesian coordinates, the orthogonal components of the intensity vector I_x, I_y and I_z can be written as:

$$\begin{aligned}
I_x &= \frac{1}{2} p_0 V_{0,x} \cos(\varphi_{\mathrm{p}} - \varphi_x),\\
I_y &= \frac{1}{2} p_0 V_{0,y} \cos(\varphi_{\mathrm{p}} - \varphi_y),\\
I_z &= \frac{1}{2} p_0 V_{0,z} \cos(\phi_{\mathrm{p}} - \phi_z).
\end{aligned} \tag{1.33}$$

Differential phase relations $\Delta\varphi_x = \varphi_{\mathrm{p}} - \varphi_x$, $\Delta\varphi_y = \varphi_{\mathrm{p}} - \varphi_y$, $\Delta\varphi_z = \varphi_{\mathrm{p}} - \varphi_z$ are crucial characteristics of the acoustic field. Notice that p and V are measured at the same point in space at which flat waves from the same source arrive by different paths. Phase differences (1.33) in the field can change significantly when the sound source moves relative to the receiver. The accompanying changes in p_0 and V_0 may be minimal.

Significant changes in the phase difference up to 2π occur because while net fluctuations of pressure add up like scalars, particle velocity oscillations add up like vectors. The full intensity vector in Cartesian coordinates:

$$\boldsymbol{I} = \boldsymbol{i} I_x + \boldsymbol{j} I_y + \boldsymbol{k} I_z \tag{1.34}$$

Since particle velocity separates into two components $\boldsymbol{V}_a$ and $\boldsymbol{V}_r$, the instantaneous intensity vector too will consist of two components: active intensity $\boldsymbol{I}(t)$ and reactive intensity $\boldsymbol{Q}(t)$:

$$\begin{aligned} Q_x &= p_0 V_{0,x} \sin \Delta\varphi_x, \\ Q_y &= p_0 V_{0,y} \sin \Delta\varphi_y \\ Q_z &= p_0 V_{0,z} \sin \Delta\varphi_z, \\ \boldsymbol{Q} &= \boldsymbol{i} Q_x + \boldsymbol{j} Q_y + \boldsymbol{k} Q_z \end{aligned} \tag{1.35}$$

As follows from (1.35), $\boldsymbol{Q} = \boldsymbol{0}$ in a single flat travelling wave and in a spherical travelling wave at $kr \gg 1$. If $\Delta\varphi_x = \Delta\varphi_y = \Delta\varphi_z = \pi/2$, $\boldsymbol{I} = \boldsymbol{0}$, and the field is purely reactive (a standing wave).

1.5 Auto- and Cross-Spectral Energy Densities

Representing energy properties of acoustic fields in terms of autospectra and cross-spectra greatly expands the possibilities of studying complex acoustic fields—especially isolating spectral components from various sources (monochromatic and noise-like) in complex interference fields of many acoustic sources and in underwater ambient noise.

Fast Fourier transform can compute spectra without first calculating autospectral or cross-spectral correlation functions, meaning that temporal realisations of a random process are directly mapped to the frequency domain. The Fourier components of random functions of time $p(t)$, $V_x(t)$, $V_y(t)$, $V_z(t)$ are defined as:

$$\begin{aligned} p_k(f, T) &= \int_0^T p_k(t) \mathrm{e}^{-\mathrm{i}2\pi f t} \mathrm{d}t, \\ V_{k,j}(f, T) &= \int_0^T V_{k,j}(t) \mathrm{e}^{-\mathrm{i}2\pi f t} \mathrm{d}t, \end{aligned} \tag{1.36}$$

where k is the number of Fourier transforms of realisations of duration T.

One-sided cross-spectral and autospectral densities are defined as

$$\begin{aligned} S_{pV_j}(f) &= \lim_{T\to\infty} \frac{2}{T} \langle p_k^*(f, T) V_{k,j}(f, T) \rangle, \\ S_{V_i V_j}(f) &= \lim_{T\to\infty} \frac{2}{T} \langle V_{k,i}^*(f, T) V_{k,j}(f, T), \quad (i \neq j) \rangle, \\ S_{V_i^2}(f) &= \lim_{T\to\infty} \frac{2}{T} \left\langle \left| V_{k,i}(f, T) \right|^2 \right\rangle, \\ S_{p^2}(f) &= \lim_{T\to\infty} \frac{2}{T} \langle |p_k(f, T)|^2 \rangle, \qquad i, j = x, y, z. \end{aligned} \tag{1.37}$$

Spectra $S_{pVi}(f)$, $S_{ViVj}(f)$, $S_{Vi}{}^2(f)$, $S_p{}^2(f)$ are identical to those calculated using correlation functions [2].

Cross-spectral densities ross-spectrum) $S_{pVi}(f)$, $S_{ViVj}(f)$ are complex quantities.

Cross-Spectrum between the acoustic pressure component and the ith component of particle velocity:

$$\begin{aligned} S_{pV_i}(f) &= C_{pV_i}(f) + iQ_{pV_i}(f) = \langle S_{V_i}(f)S_p^*(f)\rangle \\ &= \langle |p_k(f,T)|^2\rangle |S_p(f)||S_{V_i}(f)|\cos\langle\varphi_{pV_i}\rangle \\ &+ j\langle |S_p(f)||S_{V_i}(f)|\rangle \sin\langle\varphi_{pV_i}\rangle,\ (i = x, y, z)\quad j^2 = -1, \end{aligned} \tag{1.38}$$

where $C_{pV_i}(f) = |S_{pV_i}(f)|\cos\langle\varphi_{pV_i}(f)\rangle$ and $Q_{pV_i}(f) = |S_{pV_i}(f)|\sin\langle\varphi_{pV_i}(f)\rangle$ are real-valued functions.

Modulus of the $\imath$th cross-spectrum:

$$|S_{pV_i}(f)| = \left(C_{pV_i}^2(f) + Q_{pV_i}^2(f)\right)^{\frac{1}{2}} \tag{1.39}$$

Phase difference between the acoustic pressure component and the ith component of particle velocity:

$$\Delta\varphi_i(f) = \arctan\left[\frac{Q_{pV_i}(f)}{C_{pV_i}(f)}\right] = \arctan\left[\frac{\operatorname{Im} S_{pV_i}(f)}{\operatorname{Re} S_{pV_i}(f)}\right] \tag{1.40}$$

where Re and Im are the real and imaginary parts of the complex function $S_{pVi}(f)$, $i = x, y, z$.

1.6 Frequency Coherence Function

Let us write out the intensity vector as

$$\boldsymbol{I}(t, r) = \langle p(t, r)\boldsymbol{V}(t, r)\rangle_T \tag{1.41}$$

We assume that $p(t, r)$ and $\boldsymbol{V}(t, r)$ are Gaussian random functions of time t and coordinates $\boldsymbol{r}(x, y, z)$. We further assumehat the acoustic field is a stationary ergodic process with $p(t, \boldsymbol{r}) = \boldsymbol{V}(t, \boldsymbol{r}) = 0.$, i.e. acoustic pressure and velocity are centred random variables. In this case, (1.41) is a pair correlation function. Therefore, intensity is a measure of mutual space–time coherence of random variables $p(t, r)$ and $\boldsymbol{V}(t, r)\{V_x, V_y, V_z\}$ and obeys Gaussian statistics. Building on the theory of correlation coherence developed in optics and radiophysics [1], we will construct a normalised equivalent of complex intensity (1.38) in the form:

$$\gamma_i^2(f) = \frac{\left|S_{pV_i}(f)\right|^2}{S_{p^2}(f)S_{V_i^2}(f)} i = x, y, z, \quad 0 \le \gamma_i^2(f) \le 1, \tag{1.42}$$

which estimates the linear relationship between p and V in a spectral region. Random functions of time $p(t, r_0)$, $V_x(t, r_0)$, $V_y(t, r_0)$, $V_z(t, r_0)$, are measured simultaneously at one point in space, with frequency f being the running coordinate. We will call the expression (1.42) as 'complex frequency coherence function'. Similarly for the components of particle velocity

$$\gamma_{ij}^2(f) = \frac{\left|s_{V_iV_j}(f)\right|^2}{s_{V_i^2}(f)s_{V_j^2}(f)} i, j = x, y, z, \ i \ne j, \ 0 \le \gamma_{ij}^2(f) \le 1. \tag{1.43}$$

Coherence function (1.42) is the equivalent of squared normalised correlation function at a given frequency. The physical meaning of coherence function (1.42) is the square of normalised intensity of the acoustic field at a given frequency. It is more convenient than the correlation function, so for noise analysis in vector measurements we will mostly use the coherence function rather than the correlation function.

From (1.38) to (1.43) it follows that in a deterministic wave travelling along the x axis, $\gamma_x^2(f) = 1.0$, since $\Delta\varphi_x(f) = 0°$. Processes $p(t)$ and $V_x(t)$ can then be said to be coherent. In a standing wave along the x axis, processes $p(t)$ and $V_x(t)$ are also coherent and $\gamma_x^2(f) = 1.0$, since $\varphi_x(f) = 90°$.

Coherence function $\gamma_x^2(f) = 0$ if $<\cos \varphi_x(f)> \; = \; <\sin \varphi_x(f)> \; = 0$. This is only possible when processes $p(t)$ and $V_x(t)$ are out of phase, i.e. not coherent.

The coherence function may be non-zero but less than one for the following reasons: there is an element of nonlinearity between random processes $p(t)$ and $Vx(t)$; measurements contain external noise.

According to the virial theorem [4], in a stochastic acoustic field linear relationships between $p(t)$ and $V_x(t)$ must hold on average. It then follows that if $0 \le \gamma_x^2(f)$ ≤ 1, then energy flux density of the coherent component of the signal is present in the acoustic field of ambient noise. Because functions $\gamma_i^2(f)$ and $\gamma_{ij}^2(f)$ are scalars, information about the nature of the coherent component and its direction of energy transfer can be gleaned from the shape of the phase spectrum $\Delta\varphi_i(f)$ (1.40).

Using the concept of coherent output power $S_{\text{coh}}(f)$, i.e. the part of the acoustic field power described by a linear relationship between $p(t)$ and $V_x(t)$, we can write:

$$S_{\text{coh},i}(f) = \gamma_i^2(f)S_{p^2}(f), \mathrm{i} = x, y, z. \tag{1.44}$$

Then the residual spectrum associated with the incoherent diffuse component of the acoustic field will have the form:

$$S_{dif,i}(f) = \left[1 - \gamma_i^2(f)\right]S_{p^2}(f) \tag{1.45}$$

along orthogonal directions $i = x, y, z$.

From (1.42) to (1.45) it follows that the presence of diffuse (incoherent) noise reduces the coherence function but does not distort the phase (1.40).

1.7 Complex Intensity Vector

Drawing an analogy with the Umov vector, let us express the complex intensity vector $\boldsymbol{I}_\mathrm{c}$ in the form:

$$\boldsymbol{I}_\mathrm{c} = p\boldsymbol{V}^* = \boldsymbol{I} + \mathrm{i}\boldsymbol{Q} = \mathrm{Re}\,\boldsymbol{I}_\mathrm{c} + \mathrm{Im}\,\boldsymbol{I}_\mathrm{c}, \tag{1.46}$$

where p is the acoustic pressure; $\boldsymbol{V}^*$ is the complex conjugate of the particle velocity $\boldsymbol{V}$; $\boldsymbol{I}$ is the active intensity vector; $\boldsymbol{Q}$ is the reactive intensity vector; Re is the real part *of* $\boldsymbol{I}_\mathrm{c}$; and Im is its imaginary part.

Complex notation allows us to relate energy quantities to each other as follows. The complex conjugate vector $I*$ of $\boldsymbol{I}$ is

$$\boldsymbol{I}_\mathrm{c}^* = p^*\boldsymbol{V} = \boldsymbol{I} - i\,\boldsymbol{Q}, \tag{1.47}$$

where p^* is the complex conjugate of p.

$$\begin{aligned}\boldsymbol{I}_\mathrm{c} + \boldsymbol{I}_\mathrm{c}^* &= 2\boldsymbol{I}\\ \boldsymbol{I} &= \frac{\boldsymbol{I}_\mathrm{c} + \boldsymbol{I}_\mathrm{c}^*}{2} \equiv \mathrm{Re}\,\boldsymbol{I}_\mathrm{c}.\end{aligned} \tag{1.48}$$

The difference of complex conjugate intensities is the imaginary part *of* I_c:

$$\begin{aligned}\boldsymbol{I}_\mathrm{c} - \boldsymbol{I}_\mathrm{c}^* &= 2i\,\boldsymbol{Q}\\ \boldsymbol{Q}(t) &= \frac{\boldsymbol{I}_\mathrm{c} - \boldsymbol{I}_\mathrm{c}^*}{2i} \equiv \mathrm{Im}\,\boldsymbol{I}_\mathrm{c}.\end{aligned} \tag{1.49}$$

Magnitude of the complex intensity vector

$$\sqrt{\boldsymbol{I}_\mathrm{c}\cdot\boldsymbol{I}_\mathrm{c}^*} = \left[(\boldsymbol{I} + i\,\boldsymbol{Q})(\boldsymbol{I} - i\,\boldsymbol{Q})\right]^{1/2} = \left(I^2 + Q^2\right)^{1/2} = |\boldsymbol{I}_\mathrm{c}|. \tag{1.50}$$

Potential energy density is defined as

$$V = \frac{1}{2\rho c^2}\langle pp^*\rangle, \tag{1.51}$$

where < > denotes averaging over time.

Components of kinetic energy density along Cartesian axes *x*, *y*, *z*

$$\mathrm{T}_i = \frac{\rho}{2}\langle V_i V_i^* \rangle, \quad i = x, y, z, \tag{1.52}$$

where ρ is the density of the medium.

Total kinetic energy density

$$\mathrm{T} = \sum\nolimits_{i=x,y,z} \mathrm{T}_i. \tag{1.53}$$

Relationship between potential energy density and reactive intensity. It follows from (1.49) that:

$$\boldsymbol{Q} = \left(\frac{\mathrm{i}}{2}\right)[\boldsymbol{I}_c - \boldsymbol{I}_c^*] = \left(\frac{\mathrm{i}}{2}\right)[p\boldsymbol{V}^* - p^*\boldsymbol{V}],$$

applying the Euler equation, we find:

$$\boldsymbol{Q} = -\left(\frac{c^2}{\rho}\right)[p\mathbf{grad}\ p^* - p^*\mathbf{grad}\ p] = -\left(\frac{c^2}{\rho}\right)\mathbf{grad}\ V. \tag{1.54}$$

It follows from (1.54) that the reactive intensity vector is proportional to the gradient (with the opposite sign) of the potential energy, i.e. $\boldsymbol{Q}$ will decrease with increasing potential energy and vice versa. At the peaks and troughs of acoustic pressure, where $\mathbf{grad}\ U = 0$, $\boldsymbol{Q} = 0$, and $\boldsymbol{Q}$ changes its sign when it crosses over this point.

Divergence of the complex intensity vector

$$\mathrm{div}\boldsymbol{I}_c = \mathrm{div}(p\boldsymbol{V}^*) = \boldsymbol{V}^*\mathbf{grad}p + p\mathrm{div}\ \boldsymbol{V}^* \tag{1.55}$$

Applying the Euler equation and the energy conservation equation, we find:

$$\mathrm{div}I_\mathrm{c} = i\left[\frac{\omega}{2\rho c^2}p \cdot p^* - \rho\boldsymbol{V}\boldsymbol{V}^*\right] = -2i\omega[\mathrm{T} - V] = -2i\omega L, \tag{1.56}$$

where T and U are the kinetic and potential energy, respectively; $L = T - U$ is the Lagrangian function. Since (1.56) is imaginary but L is a real-valued function, then div $\boldsymbol{I} = 0$, but div $\boldsymbol{Q} = -\ 2\omega L$. It then follows that in a free field (without sources), reactive intensity has sources and drains. From (1.56) it follows that div $\boldsymbol{I}_\mathrm{c} \neq 0$ if $T - U \neq 0$. In a single flat wave propagating in infinite space, $T = U$ and thus div $\boldsymbol{I}_\mathrm{c} = 0$. In a spherical wave travelling under the same conditions, at $kr \ll 1$, as it follows from (A.10), the magnitude of particle velocity is $|V_m| = \frac{|p_m|}{\rho c \cos\phi}$, i.e. greater than

$|V_m| = \frac{|p_m|}{\rho c}$ at $kr \gg 1$. It then follows that in the near field of a spherical field at $kr \ll 1$ $T = \frac{V}{\cos^2 \phi}$, i.e. $T > U$ (at $0 < \varphi < \pi/2$) and div $\boldsymbol{Q} \neq 0$. This creates vortex transfer of sound energy in the acoustic field.

Let us now consider the vorticity of the complex intensity vector

$$\mathbf{rot}\big(p\boldsymbol{V}^*\big) = p\mathbf{rot}\ \boldsymbol{V}^* + \big[\mathbf{grad}\ p \times \boldsymbol{V}^*\big] = \big[\mathbf{grad}\ p \times \boldsymbol{V}^*\big]$$

because $\mathbf{rot}\ \boldsymbol{V}^* = 0$. Applying Euler's formula $\boldsymbol{V} = -\frac{1}{i\rho\omega}\mathbf{grad}\,p$, we find

$$\mathbf{rot}\big(p\boldsymbol{V}^*\big) = -i\omega\rho\big[\boldsymbol{V} \times \boldsymbol{V}^*\big] \tag{1.57}$$

Multiplying (1.57) by $\frac{pp^*}{pp^*}$,

$$\mathrm{rot}\big(p\mathbf{V}^*\big) = -i\omega\rho\frac{\mathbf{I}_\mathrm{c} \times \mathbf{I}_\mathrm{c}^*}{pp^*}.$$

Since the vector product $\boldsymbol{I} \times \boldsymbol{I}^*$ is

$$\boldsymbol{I}_\mathrm{c} \times \boldsymbol{I}_\mathrm{c}^* = \begin{vmatrix} & \boldsymbol{i}\,\boldsymbol{j}\,\boldsymbol{k} & \\ I_x + iQ_x & I_y + iQ_y & I_z + iQ_z \\ I_x - iQ_x & I_y + iQ_y & I_z + iQ_z \end{vmatrix} = 2i\boldsymbol{I} \times \boldsymbol{Q},$$ where $i = \sqrt{-1}$, $\boldsymbol{i}$ is a unit vector.

It then follows that

$$\mathbf{rot}\ \boldsymbol{I}_c = 2\omega\rho\frac{\boldsymbol{I} \times \boldsymbol{Q}}{pp^*},$$

but since $U = \frac{1}{2\rho c^2}pp^*$, then

$$\mathbf{rot}\ \boldsymbol{I}_\mathrm{c} = \left(\frac{\omega}{c}\right)\frac{\boldsymbol{I} \times \boldsymbol{Q}}{U}. \tag{1.58}$$

As follows from (1.58), $\mathbf{rot}\ \boldsymbol{I}c \neq 0$ with $\boldsymbol{I} \neq 0$ and $\boldsymbol{Q} \neq 0$ if vectors $\boldsymbol{I}$ and $\boldsymbol{Q}$ are non-collinear. In the case of a single spherical wave propagating in an infinite medium from a point source or a dipole field, vortices of energy exist in the near field (kr $\ll$ 1), i.e. $\mathbf{rot}\ \boldsymbol{I}c \neq 0$, but at kr $\gg$ 1 $\mathbf{rot}\ \boldsymbol{I}c \neq 0$. This means that in an infinite medium near a source, vectors $\boldsymbol{I}$ and $\boldsymbol{Q}$ can be non-collinear.

Stokes' theorem relates circulation of the energy flux density (intensity) vector along an arbitrary contour to the flux of rotation of active intensity $\mathbf{rot}(p\boldsymbol{V}^*)$ through the surface bounded by the contour, i.e. the separatrix or contour of a vortex tube.

Circulation of $\mathbf{rot}(p\boldsymbol{V}^*)$ is

$$\Gamma_\mathrm{c}(\mathbf{I}) = \oint_\mathrm{c}\big(I_x\mathrm{d}x + I_y\mathrm{d}y + I_z\mathrm{d}z\big) = \oint_\mathrm{c}\boldsymbol{I}\mathrm{d}\boldsymbol{r}. \tag{1.59}$$

By the Stokes formula, $\int_S \mathbf{rot}_n \boldsymbol{I} \mathrm{d}r = \oint_c \boldsymbol{I} \cdot \mathrm{d}\boldsymbol{r}$.

1.8 Temporal Coherence Function

The concept of coherence applies equally well to oscillations and waves of any physical nature. Coherence was first introduced in optics, followed by statistical radiophysics and physical acoustics. In vector acoustics, we use this concept to investigate the degree of coherence among four components of the acoustic field: $p(t)$, $V\{V_x(t), V_y(t), V_z(t)\}$ by means of vector intensity. In this case, it is better to investigate the coherence function rather than the correlation function.

Section 1.5 introduced the 'complex frequency coherence function' to describe the degree of coherence of a stationary wavefield at a given point in space depending on signal frequency. This function complements the autospectrum and cross-spectrum—more specifically, it indicates how coherent they are in a spectral region. We can construct a 'temporal coherence' function—that is, investigate coherence of random oscillatory processes at a chosen frequency versus time at a given point of the field. We will consider coherency properties of the acoustic field between scalar and vector quantities, or rather their projections on the coordinate axes x, y, z. Either of instantaneous intensity (Umov vector), average or complex intensity can serve as a measure of correlation. Normalised averages of these values at a point of the field versus time will be the complex temporal coherence function.

Correlation of these values is calculated for the acoustic pressure receiver and the vector receiver co-located at one point r_0 and a zero time shift τ at a frequency ω_0 between $p(r_0, \omega_0, t)$ and $\boldsymbol{V}(r_0, \omega_0, t)$. Complex intensity $\boldsymbol{I}_c(t) = p(r_0, \omega_0, t)\ \boldsymbol{V}(r_0, \omega_0, t)_T$ (where t is current time and T averaging time) is the envelope of instantaneous intensity, whose magnitude estimates the degree of correlation between $p(r_0, \omega_0, t)$ and $\boldsymbol{V}(r_0, \omega_0, t)$. Let us define at point $\boldsymbol{r}_0$ a normalised correlation function of a stationary acoustic field as follows:

$$\boldsymbol{\Gamma}(\boldsymbol{r}_0, \omega_0, t) = \frac{\langle p(t)\boldsymbol{V}^*(t)\rangle_{\mathrm{T}}}{\sqrt{\langle p(t)p^*(t)\rangle_{\mathrm{T}}\langle \boldsymbol{V}(t)\boldsymbol{V}^*(t)\rangle_{\mathrm{T}}}}. \tag{1.60}$$

$\boldsymbol{\Gamma}$ $(\boldsymbol{r}_0, \omega_0, t)$ is called the degree of complex temporal coherence. Its modulus $|\boldsymbol{\Gamma}\ (\boldsymbol{r}_0, \omega_0)|$ we call the modulus of the degree of coherence, or simply the degree of coherence.

Because $\boldsymbol{\Gamma}$ $(\boldsymbol{r}_0, \omega_0)$ is complex,

$$\boldsymbol{\Gamma}(\boldsymbol{r}_0, \omega_0, t) = \mathrm{Re}\,\boldsymbol{\Gamma}(\boldsymbol{r}_0, \omega_0, t) + i\,\mathrm{Im}\,\boldsymbol{\Gamma}(\boldsymbol{r}_0, \omega_0, t). \tag{1.61}$$

Signal processing computes three orthogonal components of the vector $\boldsymbol{\Gamma}$ $(\boldsymbol{r}_0, \omega_0, t)$:

$$\Gamma_j(\boldsymbol{r}_0, \omega_0, t) = \frac{\langle p(t)V_i^*(t)\rangle_T}{\sqrt{\langle p(t)p^*(t)\rangle_T \langle V_i(t)V_i^*(t)\rangle_T}}, \tag{1.62}$$

$$0 \le \left|\Gamma_j(\boldsymbol{r}_0, \omega_0, t)\right| \le 1, \ -1 \le \mathrm{Re}\,\Gamma_j(\boldsymbol{r}_0, \omega_0, t) \le +1,$$
$$-1 \le \mathrm{Im}\,\Gamma_j(\boldsymbol{r}_0, \omega_0, t) \le +1.0$$

The argument of the function $\Gamma_j\ (\boldsymbol{r}_0, \omega_0, t)$ is

$$\Delta\varphi_i(\boldsymbol{r}_0, \omega_0, t) = \arctan\frac{\mathrm{Im}\,\Gamma_j(\boldsymbol{r}_0, \omega_0, t)}{\mathrm{Re}\,\Gamma_j(\boldsymbol{r}_0, \omega_0, t)},$$

where $j = x, y, z$; $p(t)$ and $V_j(t)$ are analytical signals computed by Hilbert transform in the band $\Delta\omega$ with central frequency ω_0. Re $\Gamma_j(t)$ are normalised x, y and z components of intensity averaged over several periods $T = 2\pi/\omega$. Expression (1.62) is a second-order correlation coefficient with $\tau = 0$.

Higher-order coherence functions can be calculated, which carry additional information about the random wavefield.

1.9 Fourth Statistical Moment of Acoustic Intensity

Correlation theory of coherence [1, 5, 10] deals with moments higher than the second order. By considering correlation of acoustic intensity over time and at various points in space, we arrive at a fourth-order moment.

Consider instantaneous intensity vectors at two spatially separated points of the acoustic field: $\boldsymbol{I}_1(x_1, y_1, z_1, t)$ and $\boldsymbol{I}_2(x_2, y_2, z_2, t+\tau)$. We will use $\Delta\boldsymbol{I}_1(t)$ and $\Delta\boldsymbol{I}_2(t+\tau)$ to denote fluctuations of instantaneous intensity about its averages $\boldsymbol{I}_1(x_1, y_1, z_1, t)$ and $\boldsymbol{I}_2(x_2, y_2, z_2, t+\tau)$; then

$$\langle \Delta I_1(t)\Delta I_2(t+\tau)\rangle = \langle I_1(t)I_2(t+\tau)\rangle - \langle I_1(t)I_2(t+\tau)\rangle \tag{1.63}$$

According to [5], $\boldsymbol{I} = \boldsymbol{I}_1 + \boldsymbol{I}_2 + 2\sqrt{I_1}\sqrt{I_2}\mathrm{Re}\gamma_{12}(\tau)$, where $\gamma_{12}(\tau)$ is the complex degree of coherence of acoustic field at points $M_1(x_1, y_1, z_1)$ and $M_2(x_2, y_2, z_2)$ and shows to what degree the fields interact. This is determined by

$$|\gamma_{12}(\tau)|^2 = \frac{\Delta I_1(t+\tau)\Delta I_2^*(t)}{I_1 I_2}, \tag{1.64}$$

where I_1 and I_2 are average intensities at points 1 and 2, and τ is wave propagation time difference. It follows from (1.46) that normalised correlation between intensity fluctuations equals the modulus of the degree of coherence squared. It then follows that if there is a coherency relationship between two points of an acoustic field, then intensity fluctuations at these points must also be correlated. This method is used in stellar interferometry [11]. The instrument that implements this method is called the intensity interferometer [1, 11]. An underwater acoustic intensity interferometer, the equivalent of the Young–Rayleigh double-slit optical interferometer, is based on two horizontally spaced combined receivers. The slits of the interferometer are the orthogonal channels of a vector receiver with a cosine directivity pattern (see Appendix B). Full-scale experiments have shown that the outlook for the studies of space-time coherence of acoustic vector fields is a promising one [12].

1.10 Conclusions

The entire mathematical toolkit of vector acoustics is based on auto- and cross-correlation analysis of the scalar acoustic pressure and three orthogonal components of the particle velocity vector of the acoustic wave. Three orthogonal components of energy flux density vector allow the problem of signal energy movement in the real deep and shallow sea to be solved. The most important information parameter of an acoustic field are differential phase relationships among the four components of the field: $p(t)$, $V_x(t)$, $V_y(t)$ and $V_z(t)$. Signal processing methods include a multiplicative as well as an additive component. Mathematical tools used are a closed system of functions and can serve to investigate phenomena such as compensation of reciprocal energy fluxes, vortices of acoustic intensity vector, noise immunity of combined detection systems, etc.

References

1. S.M. Rytov, *Introduction to Statistical Radiophysics. Part 1. Random Processes* (Nauka, Moscow, 1976). (in Russian)
2. J.S. Bendat, A.G. Piersol, *Random Data* (Mir, Moscow, 1983). (in Russian)
3. M.A. Isakovich, *General Acoustics* (Nauka, Moscow, 1973). (in Russian)
4. L.D. Landau, E.M. Lifshitz. *Fluid Mechanics* (Nauka, Moscow, 1986); M. Born, E. Wolf. *Principles of Optics* (Nauka, Moscow, 1970). (in Russian)
5. M. Born, E. Wolf, *Principles of Optics* (Nauka, Moscow, 1970)
6. N.A. Umov. *Equations of Motion of Energy in Media.* A doctoral thesis (Odessa, 1873). (in Russian)
7. V.A. Shchurov. *Vector Acoustics of the Ocean.* (Vladivostok, Dalnauka, 2003). (in Russian)
8. J. Horton, *Fundamentals of Underwater Sound Ranging* (Sudpromgiz, Leningrad, 1961). (in Russian)
9. G.L. D'Spain, Polarization of acoustic particle motion in the ocean and relation to vector acoustic intensity, in *Proceedings of 2-nd International Workshop Acoustic Engineering and Techniques* (Harbin, China, 1999), pp. 149–164
10. S.M. Rytov, Y.A. Kravtsov, V.N. Tatarsky, *Introduction to Statistical Radiophysics. Part 2* (Nauka, Moscow, 1978). (in Russian)
11. R.K. Brown. Measurement of angular diameters of stars. Phys. Uspekhi **108**(3), 529–547 (1972). (in Russian)

12. V.A. Shchurov, E.S. Tkachenko et al., Investigating the hydroacoustic wavefield using a fourth-order statistical moment, in *8th Physics of the Geospheres National Symposium* (Dalnauka, Vladivostok, 2013), pp. 233–237

Chapter 2
Theory and Technique of Vector-Phase Underwater Acoustic Measurements

2.1 Introduction

Acoustic investigations covered by this monograph are based on simultaneous measurements at the same point of an acoustic field of the scalar acoustic pressure $p(x, y, z, t)$, the three orthogonal components of particle velocity vector $\boldsymbol{V}(x, y, z, t)$ $\{V_x, V_y, V_z\}$ and differential phase relations among them. Professor S. N. Rzhevkin coined the term 'vector-phase method' to describe this approach [1, 2].

A hydrophone, which measures pressure, is a zeroth-order acoustic transducer, and a particle velocity sensor a first-order transducer. Combining two different order transducers into one instrument creates, according to a definition by A. A. Kharkevich, a combined receiver, whose novel characteristics are different to those of its component transducers [3]. A combined hydroacoustic receiver was first mentioned in [1, 2, 4]. The author considers transducers used in Brüel and Kjær intensity probes and consisting of two closely spaced zero-order transducers (hydrophones or microphones) to be small-sized hydrophone antennas with a dipole pattern rather than combined receivers as defined in [3].

Four-component combined transducers consisting of pressure sensors and three-component particle velocity or pressure gradient sensors have been used in bottom-mounted and free-drifting combined measuring systems to study acoustic fields near the coast and in the deep open ocean. The combined receiver systems either rested on the seafloor or were anchored to the bottom and buoyed up in the water column. Data from the receiving systems were transmitted via cable or by radio to an onshore laboratory or to a research vessel. Autonomous free-drifting telemetric combined systems are designed for research in the deep open ocean and can measure at depths ranging from 20 to 1000 m. This chapter presents systems created by the Pacific Oceanological Institute from 1979 to this day.

The ultimate goal of scientific research is reliable and accurate data. The problem of jointly measuring scalar and vector quantities of the acoustic field in the real ocean environment required novel equipment and techniques for full-scale acoustic

V. A. Shchurov, *Movement of Acoustic Energy in the Ocean*,
https://doi.org/10.1007/978-981-19-1300-6_2

underwater research. Our vector acoustics equipment is an original design; these are complex acoustic measuring systems.

In his combined receivers, the author used state-of-the-art particle velocity and gradient transducers developed by the Russian Academy of Sciences, Lomonosov Moscow State University and Shtorm Design Bureau, as well as the Pacific Oceanological Institute's own designs.

2.2 Necessity and Sufficiency of the Vector-Phase Approach in Acoustics

To completely describe an acoustic field, as it follows from the equations of fluid dynamics, one needs to know eight quantities: three components of the particle velocity vector $\boldsymbol{V}(x, y, z, t)\{V_x, V_y, Vz\}$; two scalars: acoustic pressure $p(x, y, z, t)$ and medium density $\rho(x, y, z, t)$; and three differential phase relationships $\Delta\varphi_{\mathrm{p}Vx}$, $\Delta\varphi_{\mathrm{p}Vy}$, $\Delta\varphi_{\mathrm{p}Vz}$. Relationships among these quantities must be known in the process of field measurement and data processing.

Acoustic pressure $p(x, y, z, t)$ and particle velocity $\boldsymbol{V}(x, y, z, t)$ are expressed through acoustic potential $\Phi(x, y, z, t)$:

$$p(x, y, z, t) = \rho \frac{\partial \Phi(x, y, z, t)}{\partial t},$$
$$\boldsymbol{V}(x, y, z, t) = -\mathbf{grad}\ \Phi(x, y, z, t)$$

With $\Phi(x, y, z, t)$ known, pressure $p(x, y, z, t)$ and particle velocity $\boldsymbol{V}(x, y, z, t)$ can be determined at any point and at any time. Since the pressure is simply related to the potential, the direct problem can usually be reduced to finding the acoustic pressure. This approach involves operations with only the scalar pressure; we will call it the scalar approach, or scalar acoustics. In this case, solving problems that involve anisotropy of the field means using distributed arrays of pressure sensors (antennas). Usually, the distance between the neighbouring hydrophone sensors is half the wavelength of the tone signal $\lambda/2$. Low-frequency antennas ($f < 1000$ Hz, $\lambda > 1.5$ m) must reach considerable lengths, but given the limited space in which the field remains stationary and ergodic, increasing the number of hydrophones defeats the purpose. More information about the acoustic field can be gathered if one knows pressure $p(x, y, z, t)$ and the particle velocity vector $\boldsymbol{V}(x, y, z, t)$ at the same time and at one point—i.e. four characteristics of the acoustic field: p, V_x, V_y, V_z, where V_x, V_y, V_z are orthogonal components of the vector $\boldsymbol{V}$ [2, 6].

Consider the Taylor series expansion of acoustic pressure $p(x, y, z, t)$ in the vicinity of a point $M_0(x_0, y_0, z_0)$ in a small region D [6]:

$$
\begin{aligned}
p(x, y, z, t) == {} & p(x_0, y_0, z_0, t) + (x - x_0)\frac{\partial p(x_0, y_0, z_0, t)}{\partial x} \\
& + (y - y_0)\frac{\partial p(x_0, y_0, z_0, t)}{\partial y} + (z - z_0)\frac{\partial p(x_0, y_0, z_0, t)}{\partial z} \\
& + \frac{1}{2}\left\{ \begin{array}{c} (x - x_0)^2 \frac{\partial p^2(x_0, y_0, z_0, t)}{\partial x^2} + (y - y_0)^2 \frac{\partial p^2(x_0, y_0, z_0, t)}{\partial y^2} \\ +(z - z_0)^2 \frac{\partial p^2(x_0, y_0, z_0, t)}{\partial z^2} \end{array} \right\} \\
& + \left\{ \begin{array}{c} (x - x_0)(y - y_0)\frac{\partial p^2(x_0, y_0, z_0, t)}{\partial x \partial y} + (x - x_0)(z - z_0)\frac{\partial p^2(x_0, y_0, z_0, t)}{\partial x \partial z} \\ +(y - y_0)(z - z_0)\frac{\partial p^2(x_0, y_0, z_0, t)}{\partial y \partial z} \end{array} \right\}.
\end{aligned}
\tag{2.1}
$$

We will write Euler's equation $\boldsymbol{V} = -\frac{1}{\rho}\int \textbf{grad}\, p\mathrm{d}t$ for a harmonic wave of frequency ω in the form $\boldsymbol{V} = -\frac{1}{j\rho\omega}\textbf{grad}\ p$. Therefore, first derivatives $\frac{\partial p}{\partial x}, \frac{\partial p}{\partial y}, \frac{\partial p}{\partial z}$ in expansion (2.1) are proportional to the components of the particle velocity V_x, V_y, V_z. It is evident from (2.1) that by measuring the acoustic field at one point, it is possible to determine the direction to the sound source. The physical meaning of (2.1) is that we measure pressure over the entire infinitesimal region D. By measuring acoustic pressure p and components of $\textbf{grad}\,p\left(\frac{\partial p}{\partial x}, \frac{\partial p}{\partial y}, \frac{\partial p}{\partial z}\right)$ simultaneously at one point of the acoustic field and determining their phase relationships to each other, we gather complete information about the acoustic vector field at that point, provided that there are no sound sources in the region D.

In experiments, vector characteristics of the acoustic field are determined by measuring the components of pressure gradient (gradient vector sensor) or by directly measuring the orthogonal components of particle velocity (electrodynamic vector sensor). The vector-phase method is made necessary by the requirement to completely describe the acoustic field, and its adoption in the practice of acoustics is sufficient for an accurate description of the acoustic field.

2.3 Principle of Measuring the Sound Particle Velocity in an Acoustic Wave

The problem of movement of a body immersed in an ideal incompressible fluid and acted on by oscillatory motion of the fluid is considered in [7]. In the case of long wavelengths, the velocity amplitude $V(\rho)$ of a sphere of density ρ in a fluid of density ρ_0 in the field of a flat sound wave with a particle velocity amplitude of $V_0(\rho_0)$ has the form:

$$
V(\rho) = \frac{3\rho_0}{\rho_0 + 2\rho} V_0(\rho_0). \tag{2.2}
$$

It follows from (2.2) that the velocity of the sphere $V(\rho)$ relative to the particle velocity of the fluid $V_0(\rho_0)$ will depend on the sphere's density. If the sphere is denser than the fluid ($\rho > \rho_0$), then the sphere will lag behind the fluid, and $V(\rho) < V_0$; if $\rho < \rho_0$, the sphere will lead the fluid (for an air bubble, $\rho \ll \rho_0$ and $V(\rho) = 3V_0$). If the sphere is as dense as the fluid ($\rho = \rho_0$), then $V(\rho) = V_0$, i.e. the velocity of the sphere equals that of the fluid, and the fluid behaves as if the sphere does not exist. Formula (2.2) holds if the sphere's radius a is much smaller than the wavelength of sound in the fluid, while the sphere's velocity $V(\rho)$ is independent of its radius.

If wavelengths are comparable with the sphere's radius, (2.2) must be replaced with a more precise relationship. An expression was derived in [9] for the velocity of an absolutely rigid sphere by taking diffraction into account. The derivation considered only the part of the reaction of the scattered field that is due to added mass while disregarding the influence of radiation resistance, which at $ka > 1$ becomes significant. Sound scattering was calculated on the assumption that the sphere is stationary, and the effect of its oscillations on the scattering was disregarded as well.

Using a rigid sphere as a particle velocity receiver in a gas or a liquid requires a more thorough theoretical analysis of diffraction of acoustic waves with a wavelength comparable with the sphere's radius a. This problem was solved analytically in [1].

We will describe a flat wave travelling in the $+x$ direction in the form:

$$p_i = p_0 \exp\ i(\omega t - kx) \tag{2.3}$$

In what follows, we will denote the wave parameter at $r = a$ by $\alpha = \omega a/c$ and leave out the $\exp(i\omega t)$ term.

We will assume that the velocity of oscillations V arises under the action on a sphere of mass $M = \nu\rho$ of the field of an incident wave p_i and an additional pressure caused by the scattered wave and described by

$$p_s = \sum\nolimits_{m=0}^{\infty} a_m P_m(cos\theta) h_m(kr), \tag{2.4}$$

where θ is the polar angle to the x axis, $P_m(\cos\theta)$ is a Legendre polynomial and $h_m(kr)$ a spherical Hankel function of the second kind. This expression includes as yet unknown coefficients a_m. Velocity amplitude equals

$$V = \frac{F_x}{i\omega\rho V} = \frac{\int_0^{\pi} (p_i + p_s)_{r=a} \cos\theta \mathrm{d}S}{i\omega\rho\upsilon}, \tag{2.5}$$

where $\mathrm{d}S = -2\pi a^2 \mathrm{d}(\cos\theta)$, and F is the force of pressure along the x axis.

The pressure of a plane wave expanded as a series of spherical harmonics can be written in a known form [8]:

$$p_i = p_0 \int_{m=0}^{\infty} i^m (2m+1) P_m(\cos\theta) j_m^{(kr)}{}_{r=a} \tag{2.6}$$

V is found by integrating the pressure forces over the sphere:

$$V = \frac{-p_0 \sum_{m=0}^{\infty}\left\{[i^m(2m+1) + a_m h_m]2\pi a_0^2 \int_0^{\pi} P_m(cos\theta)cos\theta \mathrm{d}(cos\theta)\right\}}{i\omega\rho v}.$$

Here, functions of the argument $\alpha = ka$ are shortened to j_m and k_m. The integral equals

$$\int_0^{\pi} P_m(\cos\theta)\cos\theta d(\cos\theta) = \begin{cases} -2/3 & \text{at } m = 1 \\ 0 & \text{at } m \neq 1 \end{cases}$$

Consequently,

$$V = \frac{i3p_0 j_1 + a_1 h_1}{i\omega\rho v}. \tag{2.7}$$

Scattered pressure field on the surface of a sphere oscillating with velocity amplitude V:

$$p_s = p_s^0 + p_v = \sum_{m=0}^{\infty} a_m P_m(\cos\theta)h_m. \tag{2.8}$$

The expression for p_s^0 is known [9]:

$$p_s^0 = -p_0 \sum_{m=0}^{\infty} i^{m+1}(2m+1)P_m(\cos\theta)\sin\delta_m(\alpha)e_m^{i\delta}(\alpha)h_m(\alpha). \tag{2.9}$$

Using the method for finding the sound field of the sphere given velocity $V\cos\theta$ on its surface, we will find the coefficients b_m in the expansion of velocity potential as a series of spherical harmonics [8]. In this case, the only non-zero coefficient is

$$b_1 = \frac{V}{\mathrm{i}kD_1(\alpha)\mathrm{e}^{-\mathrm{i}\delta_1(\alpha)}} = -\frac{V}{kh_1'(\alpha)},$$

where $h_1'(\alpha) = -iD_1(\alpha)\mathrm{e}^{-\mathrm{i}\delta_1(\alpha)}$ is the derivative of the spherical Hankel function (of the second kind) with respect to its argument at $r = a$; $D_1(\alpha)$ and $\delta_1(\alpha)$ are functions introduced and tabulated in [9]. The field p_V on the sphere's surface can be found from the expression

$$p_V = i\omega\rho b_1 h_1 \cos\theta = -i\rho c\frac{h_1}{h_1'(\alpha)}\cos\theta cV \tag{2.10}$$

h_1 and h_1' are abbreviations for $h_1(\alpha)$ and $h_1'(\alpha)$.

Substituting V from (2.8) into (2.10), we find

$$p_V = -i\omega\rho\frac{h_1}{h_1'}\cos\theta\frac{3p_0 j_1 + a_1 h_1}{\mathrm{i}\omega a\rho}$$
$$= -\frac{\rho_0}{\rho}\frac{h_1}{h_1'}(i3p_0 + a_1 h_1 \cos\theta) \qquad (2.11)$$

Integrating the product of total pressure P_S and spherical function P_n over the sphere:

$$\int_0^\pi p_S P_n S = \sum_{m=0}^{\infty} a_m h_m \int_0^\pi P_m P_n \mathrm{d}S$$
$$= -p_0 \sum_{m=0}^{\infty} [i^{m=1}(2m + +1)\sin\delta_m e_m^{i\delta} h_m$$
$$\int_0^\pi -\frac{\rho_0}{\rho}\frac{\alpha h_1}{h_1'}(i3p_0 + a_1 h_1)\int_0^\pi P_n \cos\theta \mathrm{d}S].$$

Considering that

$$\int_0^\pi P_m P_n \mathrm{d}_1 S = \begin{cases} -2\pi a^2 \frac{2}{2m+1}; & n = m \text{ and } \cos\theta = P_1 \\ 0 & n \neq m \end{cases}$$

and the equation given on page 212 in [8]

$$j_1' = -D_1 \sin\delta_1; \quad \sin\delta_1 \mathrm{e}^{i\delta_1} = i\frac{h_1}{h_1'}, \qquad (2.12)$$

we arrive at separate equations for each $n = m$, which can be solved for coefficients a_m:

$$a_j = i3p_0 \frac{\frac{\rho_0}{\rho}\frac{i_1}{\alpha} - j_1'}{h_1' - \frac{\rho_0}{\rho}\frac{h_1}{\alpha}} \quad \text{at } m = 1 \qquad (2.13)$$

$$a_m = -p_0 i^{m+1}(2m + 1)\sin\delta_m \mathrm{e}_m^{i\delta} h_m \quad \text{at } m \neq 1 \qquad (2.13\text{a})$$

Expression (2.13) gives the expansion coefficient in (2.4) for a scattered wave of order $m = 1$ taking into account vibration of the sphere acted on by the incident wave. Expression (2.13a) for $m \neq 1$ coincides with (2.9) for the scattering coefficient on a stationary sphere, which means that the resulting oscillations do not generate spherical waves of an order other than $m = 1$.

Substituting a_1 in (2.7), we find

$$V = \frac{3p_0}{\omega a\rho}\left[j_1 + h_1\frac{\frac{\rho_0}{\rho}\frac{i_1}{\alpha} - j_1'}{h_1' - \frac{\rho_0}{\rho}\frac{h_1}{\alpha}}\right] = \frac{3p_0}{\omega a h_1'}\frac{j_1 h_1' - h_1 j_1'}{\rho - \rho_0\frac{h_1}{\alpha h_1'}}. \tag{2.14}$$

Using the expression for spherical Bessel functions and their derivatives, we will write

$$j_1 h_1' - h_1 j_1' = \frac{1}{i\alpha^2}, \tag{2.15}$$

and also find the following ratio:

$$\frac{h_1(\alpha)}{h_1'(\alpha)} = -\frac{\alpha}{2}\left[2\frac{2+\alpha^2}{4+\alpha^4} - \mathrm{i}\frac{2\alpha^3}{4+\alpha^4}\right] = \frac{\mathrm{i}Z_1}{\frac{1}{3}S\rho c} = -\frac{\alpha}{2}\mu, \tag{2.16}$$

where

$$Z_1 = \frac{1}{3}S\rho c\frac{\alpha^4}{4+\alpha^4} + \mathrm{i}\omega\rho V + \frac{2+\alpha^2}{4+\alpha^4} \tag{2.17}$$

is the impedance of an oscillating sphere, $S = 4\pi a^2$, $V = 4/3\pi a^3$, and

$$\mu = 2\left[\frac{2+\alpha^2}{4+\alpha^4} - i\frac{2\alpha^3}{4+\alpha^4}\right]. \tag{2.18}$$

At $\alpha \ll 1$, μ tends to unity.

Using these expressions, we obtain for the complex velocity amplitude of the sphere

$$V(\alpha) = \frac{\frac{3}{2}\rho e^{i\delta_1(\alpha)}V_0}{\left(\rho + \frac{\rho_0\mu}{2}\right)\sqrt{1+\frac{\alpha^4}{4}}} = \frac{3\rho_0}{\left(\rho + \frac{\rho_0\mu}{2}\right)}\frac{e^{i\delta_1(\alpha)}}{\sqrt{1+\frac{\alpha^4}{4}}}V_0. \tag{2.19}$$

In the long wavelength limit ($\alpha \ll 1$), $V(\alpha)$ becomes (2.2).

We get the same result as (2.18) if in deriving the pressure force F_x on a stationary sphere we consider only the pressure $\left(p_i^0 + p_S^0\right)|_{r=a}$ on a stationary sphere [8]:

$$F_x = \frac{4\pi a^2 e^{i\delta_1(\alpha)}}{\alpha^2 D_1(\alpha)}p_0, \quad \text{where} \quad D_1(\alpha) = \sqrt{\frac{4+\alpha^4}{\alpha^3}},$$

but to find the velocity amplitude, we need to augment the sphere's inertial impedance $i\omega\rho v$ by impedance Z_1 (2.17) to describe reaction of ambient field. Therefore, it doesn't matter whether we consider the additional pressure forces p_V arising from oscillations of the sphere with velocity amplitude V or find the total force acting on the sphere and disregard p_V, but we do need to add to the sphere's own impedance the impedance caused by reaction of radiation field (2.17).

Both pathways are incorrect to a degree, since they use expressions for pressure forces that arise when sound is scattered on a stationary sphere while in fact the sphere is oscillating. We should note, however, that the amplitude of the sphere's oscillations is very small compared with the wavelength, so the boundary condition on a stationary sphere will not noticeably differ from that on a moving sphere.

Figure 2.1 charts the sphere's velocity amplitude as a function of $\alpha = \omega a/c = ka$ at different relative densities of the medium ρ_0 and average density of the sphere ρ; the charted value is the ratio of velocity amplitude $V(\alpha)$ to velocity amplitude (2.2) in an incompressible liquid (1—$\rho = 0.5\ \rho_0$; 2—$\rho = 1.0\ \rho_0$; 3—$\rho = 2.7\ \rho_0$).

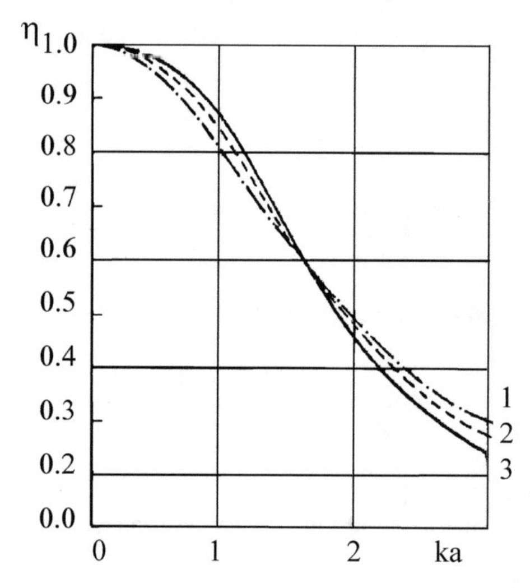

Fig. 2.1 Dependence of $\eta = |V(ka)|/V(\rho)$ on wave parameter ka for spherical bodies of various densities ρ. 1—$\rho = 0.5\ \rho_0$; 2—$\rho = 1.0\ \rho_0$; 3—$\rho = 2.7\ \rho_0$ [1]

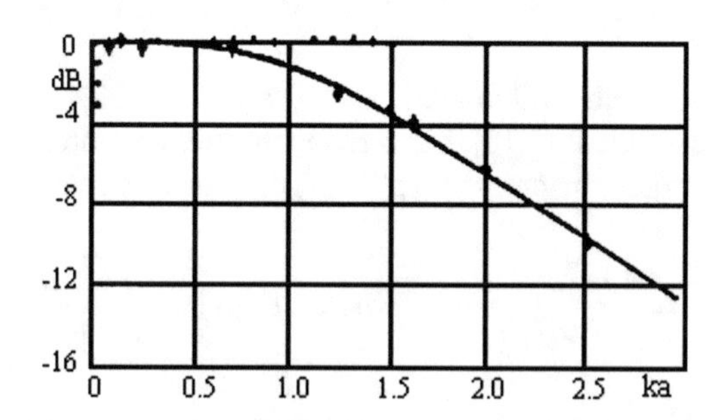

Fig. 2.2 Velocity amplitude $V(ka)$ (solid line); dots are experimental data for a spherical body of density $\rho = 2.7\rho_0$ [1]

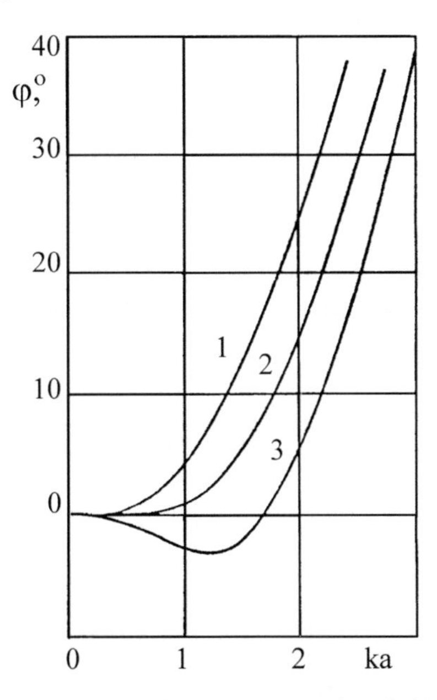

Fig. 2.3 Phase difference $\varphi(ka)$ between the velocity of a spherical body of density ρ and the particle velocity in an incoming flat wave in a fluid of density ρ_0. 1—$\rho = 0.5\ \rho_0$; 2—$\rho = 1.0\ \rho_0$; 3—$\rho = 2.7\ \rho_0$ [1]

Figure 2.2 shows velocity amplitude $V(\alpha)$ for the case of $\rho = 2.7$ and experimental points from calibration of a spherical receiver in water. Figure 2.3 shows phase difference φ between the particle velocity and the phase of oscillations in the incident wave (same notation as in Fig. 2.1).

We can also derive the amplitude of oscillation velocity of a sphere acted on by a sound wave from the formula for scattering of sound on a flexible sphere from [8]. We will assume that the sphere is very rigid but its density is of the same order of magnitude as the ambient fluid; this means that the speed of sound in the material of the sphere is $c \gg c_0$. Under these conditions, pulsations of the sphere will obviously be very weak while oscillations will be finite.

For the amplitude of oscillations of a sphere of density ρ in a fluid of density ρ_0, using the boundary conditions that sound pressure and normal component of velocity must be equal inside and outside the sphere at $r = a$, we find

$$\begin{aligned} V &= -\frac{k}{i\omega\rho}\left\{i3p_0[-D_1(\alpha)\sin\delta_1(\alpha)] + A_1 D_1(\alpha)e^{-\left[i\delta_1(\alpha)+\frac{\pi}{2}\right]}\right\} \\ &= \frac{3p_0}{\rho_0 c}j_1' - iA_1\frac{h_1'}{\rho_0 c}, \end{aligned} \tag{2.20}$$

in which the first term is due to the incident wave and the second to the scattered wave. Coefficient A_1, which determines the amplitude of the first-order scattered wave, is given by the formula

$$A_1 = -3p_0 \frac{\frac{\overline{\rho c}}{\rho_0 c_0} D_1 \sin\delta_1 \overline{j_1} - \overline{D_1} \sin\overline{\delta_1}\, j_1}{\frac{\overline{\rho c}}{\rho_0 c_0} D_1 e^{-i\delta_1 \overline{j_1}} + iG_1 e^{-i\varepsilon_1} \overline{D_1} \sin\overline{\delta_1}}, \tag{2.21}$$

where $G_1 \mathrm{e}^{-i\varepsilon_1} = h_1$; functions without the overbar are functions of argument α, and the overbar means that the function's argument is $\alpha' = \frac{\omega a}{c} = \frac{c_0}{c}\alpha$. Recasting the functions in (2.21) as follows:

$$\overline{j}_1 = \frac{c_0}{c} j_1 \frac{\overline{j}_1 c}{j_1 c_0} = \frac{c_0}{c} j_1 \beta_1 \quad \text{and} \quad \overline{j}'_1 = j'_1 \frac{\overline{j}'_1}{j'_1} = j'_1 \beta_1,$$

Equation (2.21) takes the form

$$A_1 = \mathrm{i}3p_0 \frac{\frac{\rho}{\rho_0} j'_1 j_1 \beta_1 - j'_1 j_1 \beta_2}{\frac{\rho}{\rho_0} h_1 j_1 \beta_1 - h_1 j'_1 \beta_2}. \tag{2.22}$$

Substituting this in (2.20), we find

$$V = \frac{3p_0}{\rho c} \frac{j'_1 \beta_2}{j_1 \beta_1} \frac{j_1 h'_1 - h_1 j'_1}{h'_1 \frac{\rho}{\rho_0} - \frac{h_1 j'_1 \beta_2}{h'_1 j_1 \beta_1}}.$$

Using (2.15) and (2.16), we finally find for the sphere velocity amplitude:

$$V = \frac{\frac{3}{2}\rho_0}{\rho + \frac{\rho_0}{2}\gamma(\overline{\alpha})\mu(\alpha)} \frac{\gamma(\overline{\alpha})\mathrm{e}^{i\delta_1(\alpha)}}{\sqrt{1 + \frac{\alpha^4}{4}}} V_0, \tag{2.23}$$

where,

$$\gamma(\overline{\alpha}) = \frac{j'_1(\alpha)}{j_1(\alpha)} \alpha \frac{\beta_2}{\beta_1} = \frac{\overline{j}'_1}{\overline{j}_1} \overline{\alpha} = -\frac{\overline{D}_1 \sin\delta_1}{\overline{j}_1(\overline{\alpha})}. \tag{2.24}$$

Multiplier $\gamma(\overline{\alpha})$ can be taken from tables for $D_1(\overline{\alpha})$ and $\delta_1(\overline{\alpha})$ in [9]. For $\alpha = 0.4$; 0,6; 1.0 we find $\gamma(\overline{\alpha}) = 0.96$; 0.89; 0.81, respectively. The corresponding values of $\alpha = \frac{\overline{c}}{c}\overline{\alpha}$ are found by multiplying by the ratio of velocity in the solid sphere to velocity in the fluid (approximately three). Therefore, when $\alpha \leq 1$, i.e. $\lambda \geq 2a$, we find $\overline{\alpha} \leq 0.3$, the correction factor γ approaches unity, and (2.23) yields values close to (2.18) derived for an absolutely rigid sphere, for which $\overline{c} = \infty$ and $\gamma(\overline{\alpha})$ is exactly one.

There is another way to determine the oscillation velocity of a spherical body in a fluid: using conservation of momentum. The momentum of a spherical body (of volume v and density ρ) oscillating with velocity amplitude V plus the total momentum of fluid oscillating under the action of the body moving in it (with a relative velocity) must equal the momentum of volume v cut out of the fluid and oscillating under the action of the flat wave (2.3) with the velocity amplitude that existed before the body was introduced.

The velocity amplitude of this volume equals

$$V = \frac{\int_0^\pi p_i|_{r=a} \cos \upsilon \mathrm{d}S}{\mathrm{i}\omega\rho\upsilon} = \frac{\mathrm{i}4\pi a^2 3 j_1}{\mathrm{i}\omega\rho\frac{4}{3}\pi a^3} p_0 = \frac{3 j_1}{\alpha} V_0, \tag{2.25}$$

where p_i is found from (2.6) at $\alpha \ll 1$ $V \approx V_0$.

The momentum of the spherical volume v cut out of the fluid equals

$$\frac{3 j_1}{\alpha} \rho \upsilon V_0. \tag{2.26}$$

The (radial) velocity of fluid particles on the surface of the sphere in wave motion (2.3) is a negative derivative of the velocity potential with respect to the radius at $r = a$

$$V' = -\left.\frac{\partial \Phi_1}{\partial r}\right|_{r=a} = \left.\frac{-k}{i\omega\rho} \frac{\partial p_i}{\partial (kr)}\right|_{r=a} = \frac{i p_0}{\rho c} \sum_{m=0}^{\infty} i^m (2m+1) P_m(\upsilon) j'_m,$$

where p_i is an infinite sum (2.6).

In the resulting expression, we need only to keep the term with $m = 1$, which describes oscillation of the spherical volume along the x axis, i.e.

$$V' = \frac{i}{\rho c}\left[\mathrm{i}3 \cos\theta \overline{j}'_1\right] p_0 = -3\overline{j}'_1 \cos\theta V_0.$$

At $\theta = 0$ (towards the incident wave), we find velocity $V'_0 = -3\overline{j}'_1 V_0$ along the radius, i.e. in the $+x$ direction at $\theta = \pi$ the directional velocity has the same magnitude with a minus sign, i.e. it is also oriented along the $+x$ axis. Accordingly, the velocity amplitude along the x axis of the volume of fluid v as a whole due to the sound wave equals

$$V' = 3 j'_1 V_0 \tag{2.27}$$

at $\alpha \ll 1$, $V \approx V_0$.

The velocity of the body relative to the surrounding fluid is (in amplitude) $V - V' = \upsilon - 3 j'_1 V_0$, and the momentum of the fluid caused by the motion is the product of relative velocity and added mass. The added mass M' according to (2.16) can be expressed through impedance Z_1 of the oscillating sphere

$$M' = \frac{Z_1}{i\omega} = V\frac{p}{2}\mu.$$

Hence the momentum of the surrounding fluid together with the momentum of the moving body will equal (in amplitude)

$$M' = v\frac{p}{2}\mu\left[V - 3j_1'V_0\right] + \rho v V. \tag{2.28}$$

Equating this value with the momentum of the imaginary spherical volume cut out of the fluid (2.26), we get an equation from which we find

$$V = 3\rho\frac{\frac{j_1}{\alpha} + j_1'\frac{\mu}{2}}{\rho + \frac{\rho}{2}\mu}V_0.$$

Given that according to (2.16) $\frac{\mu}{2} = -\frac{h_1}{\alpha h_1}$, after transformations we find

$$V = \frac{\frac{3}{2}\rho e^{i\delta_1(\alpha)}V_0}{\left(\rho + \frac{\rho\mu}{2}\right)\sqrt{1 + \frac{\alpha^4}{4}}},$$

i.e. exactly the same expression as (2.18).

The same method solves the problem of oscillation in the fluid of density ρ_0 of an infinite round cylinder (Fig. 2.4) of density ρ acted on by a flat wave travelling in the $+x$ direction perpendicular to the cylinder's axis. Borrowing from [8] the expression for the total pressure of incident and scattered waves on the surface of a stationary

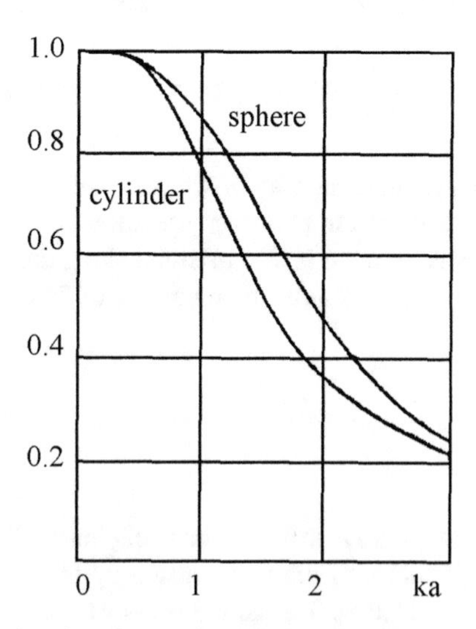

Fig. 2.4 Velocity amplitudes of a sphere and a cylinder versus (ka) at $\rho/\rho_0 = 2.7$

cylinder (of radius a), we find

$$p_i(r) + p_S(r)|_{r=a} = \frac{4p_0}{\pi\alpha}\sum_{m=0}^{\infty}\frac{\cos m\varphi}{C_m(\alpha)}e^{i\left[\gamma_m(\alpha)-\frac{m\pi}{2}\right]}, \tag{2.29}$$

where $C_m(\alpha)$ and $\gamma_m(\alpha)$ are auxiliary functions tabulated in [8].

To find the force acting on the cylinder in the x direction per unit length of the cylinder:

$$\begin{aligned} F_x &= \int_0^{2\pi}[p_i + p_S]_{r=a}a\cos\varphi\mathrm{d}\varphi \\ &= \frac{4a\rho_0 c}{\pi\alpha}v_0\sum_{m=0}^{\infty}\frac{e^{i\left[\gamma_m(\alpha)-\frac{m\pi}{2}\right]}}{C_m(\alpha)}\int_0^{2\pi}\cos m\varphi\cos\varphi\mathrm{d}\varphi. \end{aligned}$$

The integral in this expression is non-zero only at $m = 1$ and equals π. Then,

$$F_x = i\frac{4a\rho_0 c e^{i\gamma(\alpha)}}{\alpha C_1(\alpha)}V_0. \tag{2.30}$$

The (complex) amplitude of the particle velocity equals

$$V = \frac{F_x}{i\omega\pi a^2\rho + Z_1'}, \tag{2.31}$$

where Z_1' is impedance per unit length of a cylinder oscillating along the x axis with a velocity amplitude v perpendicular to its axis.

According to the derivation in [8], we will write

$$\begin{aligned} Z_1' &= \pi a\rho_0 c\frac{H_1(\alpha)e^{i\gamma_1(\alpha)}}{C_1(\alpha)} \\ &= -i\omega\pi a^2\rho_0\frac{H_1(\alpha)}{\alpha H_1'(\alpha)} = i\omega\pi a^2\rho\mu'(\alpha), \end{aligned} \tag{2.32}$$

where $H_1(\alpha) = I_1(\alpha) – iN_1(\alpha)$ is a Hankel function of the second kind;

$$H_1'(\alpha) = \left.\frac{\mathrm{d}H(kr)}{\mathrm{d}(kr)}\right|_{r=a} = -iC_i(\alpha)e^{-i\gamma_1(\alpha)}. \tag{2.33}$$

In (2.32) we introduced the variable

$$\mu' = -\frac{H_1(\alpha)}{\alpha H_1'(\alpha)} = \frac{I_1(\alpha) - iN_1(\alpha)}{i\alpha C_1(\alpha)}e^{i\gamma_1(\alpha)}. \tag{2.34}$$

Substituting Z_1' in (2.31), we find

$$V = \frac{4\rho e^{i\gamma_1(\alpha)} V_0}{\pi a^2 C_1(\alpha)(\rho + \rho_0 \mu')} = \frac{2\rho_0}{(\rho + \rho_0 \mu)} \left[\frac{2e^{i\gamma_1(\alpha)}}{\pi a^2 C_1(\alpha)} \right] V_0. \quad (2.35)$$

At $\alpha \ll 1$, using limit values $C_1(\alpha) \approx 2/\pi\alpha^2$; $\gamma_1(\alpha) \approx 0$; $\mu \approx 1$, we satisfy ourselves that the bracketed term tends to unity.

We then arrive at

$$V = \frac{2\rho_0}{\rho_0 + \rho} V_0, \quad (2.36)$$

derived in incompressible fluid dynamics [7].

Our theoretical analysis leads to these conclusions. The force that causes the sphere to move in the fluid is proportional to the pressure gradient in the acoustic wave. Formula (2.1) holds if $ka < 1$, where k is the wavenumber and a the radius of the sphere. It then follows that $a < \lambda/2\pi \leq \lambda/6$, i.e. the diameter of the sphere must meet the condition $D = 2\alpha < \lambda/3$.

Relation (2.1) forms the basis of the instrument for measuring the acoustics particle velocity—the vector receiver.

2.4 Vector Acoustic Receiver

A spherical or cylindrical body co-oscillating with the particles of the liquid or gas can be used to build an instrument to take point measurements of orthogonal components of vector quantities such as particle displacement $\boldsymbol{\xi}(t)\{\xi_x(t), \xi_y(t), \xi_z(t)\}$, particle velocity $\boldsymbol{V}(t)\left\{\frac{\partial \xi_x(t)}{\partial t}, \frac{\partial \xi_y(t)}{\partial t}, \frac{\partial \xi_z(t)}{\partial t}\right\}$, particle acceleration $\boldsymbol{a}(t)\left\{\frac{\partial^2 \xi_x(t)}{\partial^2 t}, \frac{\partial^2 \xi_y(t)}{\partial^2 t}, \frac{\partial^2 \xi_z(t)}{\partial^2 t}\right\}$ and acoustic pressure gradient **grad** $p(t)\left\{\frac{\partial p(t)}{\partial x}, \frac{\partial p(t)}{\partial y}, \frac{\partial p(t)}{\partial z}\right\}$.

Placing electroacoustic transducers of acoustic particle displacement, velocity or acceleration along three orthogonal Cartesian axes x, y, z inside a spherical body creates a three-component electroacoustic transducer commonly called a vector receiver [2].

In his research, the author used two types of vector receivers developed in the USSR. In the 1–100 Hz frequency range, we used an electrodynamic particle velocity sensor (Fig. 2.5); and in the frequency range of 10–1000 Hz, a piezoelectric pressure gradient sensor (Fig. 2.6).

Interestingly, a device was invented in ancient China 2000 years ago to determine the direction to the centre of an earthquake, and according to Chinese chronicles, an array of such devices could determine the distance to the centre of an earthquake. As Fig. 2.7 illustrates, the movement of the device's massive body coupled to the Earth's surface relative to the pendulum ejected a copper ball from the dragon's

Fig. 2.5 Electrodynamic low-frequency three-component vector receiver. Maximum deployment depth: 1000 m. Operating range: 1–100 Hz. Axial sensitivity of channels: 10 μV/Pa

Fig. 2.6 Acoustic low-frequency four-component combined receiver in the housing basket. The vector receiver (sphere) is a three-component pressure gradient receiver. Operating frequency range: 10–1000 Hz. Pressure transducer sensitivity: 500 μV/Pa. Axial sensitivity of the vector channel: 1200 μV/Pa at a frequency of 1000 Hz

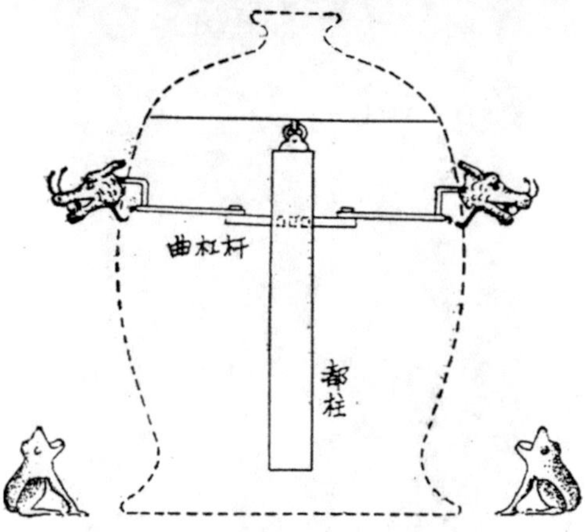

Fig. 2.7 This seismic instrument is 2000 years old. A replica is exhibited in Taipei (Taiwan) at the National Museum of Natural Sciences

mouth. Angle 2π was divided into equal angular sectors $\pi/4$, which apparently was enough to estimate the bearing. We should note that the same principle is used in today's co-oscillating inertial receivers.

2.4.1 Basic Specifications for a Vector Receiver

Specifications for a vector receiver include: directivity pattern of each individual transducer; frequency characteristic of sensitivity of each individual transducer; differential phase characteristics between transducers of orthogonal axes *x, y, z*.

Each of the three identical transducers of the vector receiver has one degree of freedom and can register longitudinal oscillations only along one of the orthogonal axes *x, y, z* As noted above, the directivity pattern of such a transducer about its axis has the form of $\cos\theta$, where θ is the angle between the transducer axis and the given direction. This directivity pattern is called a dipole pattern. In space, the directivity pattern of an ideal dipole transducer appears as two touching spheres. A tangent plane through the point of contact of these spheres is the transducer's plane of zero sensitivity.

A line perpendicular to that plane and passing through the point of contact of the spheres is the transducer's axis of maximum sensitivity. Figure 2.8 illustrates the directivity pattern of an ideal dipole transducer. The orthogonal transducers that make up a vector receiver must have identical directivity patterns and sensitivities.

The ratio of axial (maximum) sensitivity to transverse sensitivity of a real transducer is called the sensitivity relation coefficient. The best directivity patterns of real receivers have a sensitivity relation coefficient of 28–30 dB. The directivity pattern of a dipole transducer is independent of frequency in the operating frequency range of a real transducer.

In our studies, the operating frequency range went up to 1000 Hz, which translates to the minimum wavelength of $\lambda_{\min} \approx 1.5\ m$. Diameters of vector receivers used in the author's research met the condition $D \leq 0.2$ m. This means that the condition for $\lambda_{\min}$, $D \leq \lambda \leq \lambda_{\min}/3 \approx 0.5$ m, was met [2].

In a vector receiver, ideal transducers of the orthogonal Cartesian axes *x, y, z* have the following spatial directivity patterns in spherical coordinates (R, φ, θ):

$$R_x = R_0 \sin\theta \cos\phi, \quad R_y = R_0 \sin\theta \sin\phi, \quad R_z = R_0 \cos\theta, \tag{2.37}$$

where R_0 is the axial sensitivity of dipole transducers of channels *x, y, z;* φ is the azimuthal angle measured from the *x* axis; θ is the polar angle measured from the *z* axis.

With the three orthogonal channels (*x, y, z*) considered together, the directivity pattern of the vector receiver is a sphere:

$$R_x^2 + R_y^2 + R_z^2 = R_0^2 \tag{2.38}$$

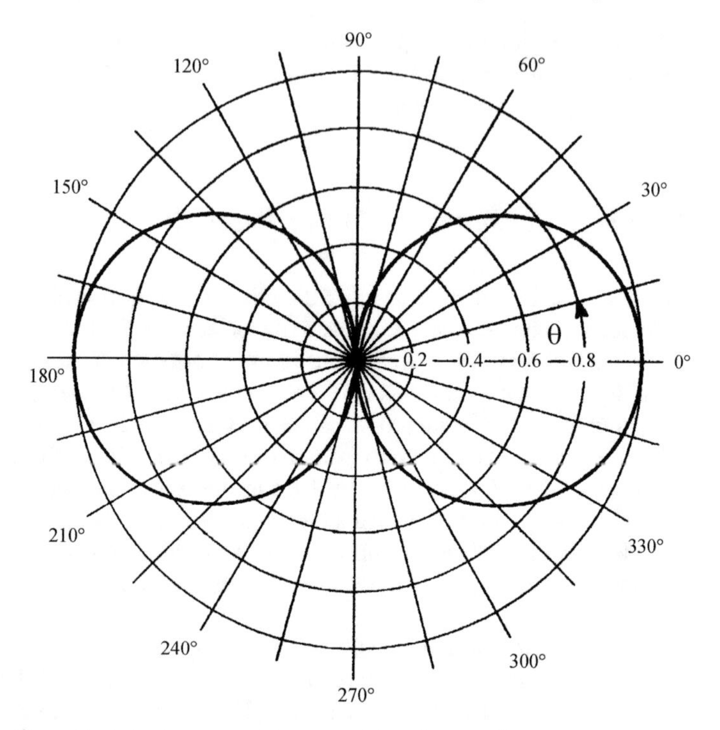

Fig. 2.8 Normalised dipole pattern $V/V_0 = \cos\theta$ in polar coordinates on a linear scale. Legend: V_0 is sensitivity at $\theta = 0°$, and V at $\theta \neq 0°$

2.4.2 *Piezoceramic and Electrodynamic Vector Receivers*

Expression (2.2) suggests that positive buoyancy increases and negative buoyancy decreases a vector receiver's sensitivity. Sensitivity of a vector receiver (or a hydrophone, for that matter) is usually defined as its emf divided by the sound pressure. Frequency characteristic means sensitivity of a transducer as a function of frequency. Because each channel of the vector receiver has a sensitivity that varies with direction as $V = V_0\cos\theta$, we will restrict our discussion to axial sensitivities of each channel *x, y, z*. Sensitivity of the receiver varies with the principle of transduction used. In a piezoelectric transducer, inertial masses of sensors exert a force $f = -ma$ on a piezoceramic element, where m is the inertial mass and a the acceleration of inertial mass relative to the receiver body. The piezoceramic element behaves like a lightweight rigid spring whose natural frequency lies above the operating frequency range. In this case, the velocities of all internals and the receiver's body are the same and proportional to the particle velocity in a flat wave. The ratio of output voltage *e* of a piezoceramic transducer to pressure p is proportional to frequency ω, i.e. $e/p \propto \omega$. As a result, voltage sensitivity in the free field increases with frequency at a rate of 6 dB/octave [28].

In a moving coil (electrodynamic) sensor, the coil moves in the air gap of a permanent magnet. The emf in the coil is proportional to the relative velocity of the coil and the magnet. The coil and the magnet are weakly linked mechanically, so the system has a low natural frequency. The receiver's operating range lies above the natural frequency of the sensor, which is why the freely suspended part of the sensor remains almost stationary in space, while the part attached to the sphere moves at a speed equal to the particle velocity.

Axial sensitivity of the particle velocity sensor is independent of frequency, i.e. it is a constant. One feature of such a transducer is its low internal resistance, and thus low self-noise. Differential phase relations $\Delta\varphi_{xy}$, $\Delta\varphi_{xx}$, $\Delta\varphi_{yz}$ depend on the quality and sameness of vector channels *x, y, z*. In a flat travelling wave, they must equal 0° or 180°.

A particularly important feature of transducers is their bandwidth. The theoretical dependence of sensitivity on frequency is an approximation, since receiver body imbalance and other imperfections of real receivers may introduce parasitic resonances. These must be corrected during their manufacture and adjustment. This is why real vector receivers are described in terms of mean sensitivity. It is measured in $\mu V/Ra$ at a mean frequency $f_{\mathrm{mean}} = (f_{\mathrm{u}} + f_{\mathrm{l}})/2$ or $(f_{\mathrm{u}} \cdot f_{\mathrm{l}})^{1/2}$, where f_{u} and f_{l} are the upper and the lower end of the receiver's operating frequency range. The mean sensitivity is the sensitivity at frequency f_{mean} linearly interpolated from measured real sensitivity between f_{u} and f_{l}. Sensitivity is usually plotted in log–log scale. The slope of the resulting straight line is the frequency dependence of mean sensitivity.

A crucial technological feature of the vector receiver is a common phase centre for all channels, which must be in the centre of the sphere, with the *x, y, z* channel transducers arranged symmetrically about it. The centres of gravity and buoyancy must also be in the geometric centre of the receiver sphere. To learn more about the various types of vector receivers, the reader is referred to [19].

2.5 Combined Acoustic Receiver

The combined four-component hydroacoustic receiver is a measuring instrument consisting of a hydrophone—a zero-order (scalar) transducer—and a vector receiver consisting of three first-order transducers.

The combined receiver measures four physical quantities simultaneously at one point of the acoustic field: acoustic pressure *p*(*t*) and three orthogonal components of the particle velocity vector $\boldsymbol{V}(t)\left\{V_x(t),\ V_y(t),\ V_z(t)\right\}$ or pressure gradient vector $\mathbf{grad}\ p(t)\left\{\frac{\mathrm{d}p(t)}{\mathrm{d}x}, \frac{\mathrm{d}p(t)}{\mathrm{d}y}, \frac{\mathrm{d}p(t)}{\mathrm{d}z}\right\}$. For simplicity, we will call this a four-channel receiver and denote the channels of the combined receiver *p, x, y, z*. An ideal four-component combined receiver must have a spherical directivity pattern: sensitivity of the p channel and its phase response must be frequency-independent in the operating frequency range; the x, y, z channels of the vector receiver must have a dipole directivity pattern and ideal phase response. Channels p, x, y, z must share a phase centre. In addition, the centre of gravity, the buoyancy centre and the phase centre of the combined receiver must all be at the same point—the vector receiver's geometric centre. When these conditions are met, the directivity pattern of the combined receiver is a sphere, i.e. identical to that of the vector receiver (2.37), (2.38).

The dipole directivity pattern of each of the vector channels x, y, z is bidirectional (Fig. 2.8). In a combined receiver, it is possible to create a unidirectional additive directivity pattern—a cardioid. Figure 2.9 shows a normalised directivity pattern $p/p_0 = 1/2(1 + \cos\theta)$, which is the sum of sensitivities of a single channel x dipole transducer and the omnidirectional hydrophone p channel. To produce this pattern, axial sensitivity of the dipole receiver is matched to the sensitivity of the omnidirectional hydrophone. This directivity pattern is called a cardioid. The '+' sign corresponds to zero phase difference between the hydrophone channel and the single channel of the vector receiver. Cardioid of the form $p/p_0 = 1/2(1 - \cos\theta)$ corresponds to a 180° phase difference. '+' and '–' cardioids are rotated 180° relative to each other (antiphase cardioids).

Although dipole and cardioid directivity patterns of the vector receiver have a directivity factor of three, they are widely used in low-frequency applications [10].

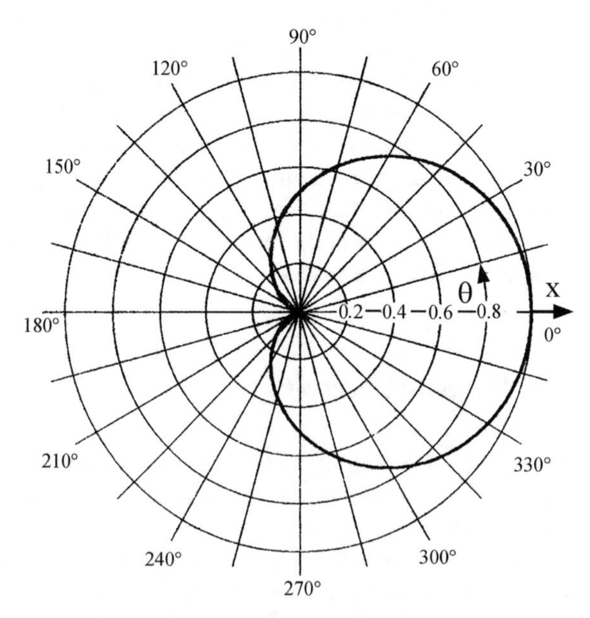

Fig. 2.9 Cardioid directivity pattern in polar coordinates on a linear scale, given by the formula $p/p_0 = 1/2(1 + \cos\theta)$: sensitivity is p_0 at $\theta = 0°$, p at $\theta \neq 0°$. The angle θ is measured from the x axis

With four channels (*p, x, y, z*) in combinations *p* and *x*, *p* and *y*, *p* and *z*, there can be six unidirectional channels. We will denote them $p \pm x$, $p \pm y$, $p \pm z$, which gives us three pairs of cardioids along the *x, y, z* axes. Each of the channels *x, y, z* will consist of two antiphase cardioids. Such directivity patterns are mostly used in applications. A study of underwater ambient noise using cardioid directivity patterns is published in [5, 10–12].

With a diverse set of directivity characteristics described above, the combined receiver is nevertheless a point receiving system for the frequency range in which its linear geometric dimensions do not exceed 1/3 of the wavelength at the upper end of the operating frequency range.

Figure 2.6 shows an assembled combined acoustic receiver. Attached to the spherical body of the vector receiver are six hydrophones arranged symmetrically about the geometric centre of the sphere and the phase centre. With the six hydrophones arranged this way, there is no geometric phase advance between *p* and V_x, V_y, V_z. Experiment has shown that with fewer than six hydrophones, the combined receiver cannot be used as a physical instrument [5, 17, 21].

The quality of a real combined receiver depends on the technical characteristics of all its four channels as well as on how well the vector receiver is suspended and how the hydrophone and the vector receiver are arranged in the instrument module. These matters have been discussed by many researchers. The results can be summarised as follows [13–19].

Vector receivers usually have an average density that is either below or just above the sea water density. Designing a neutrally buoyant receiver ($\rho = \rho_0$) is challenging from an engineering viewpoint. Furthermore, attaching and orienting such a receiver inside the instrument module is also quite difficult. Vector receivers used by the author had an average density of $\rho \approx 0.7$–0.8 g/cm^3 ($\rho < \rho_0$) and $\rho \approx 1.2$–1.5 g/cm^3 ($\rho > \rho_0$).

The vector receiver must be suspended inside the instrument module in such a way that one of the receiver's orthogonal channels (usually the *z* channel) would default to a vertical position when in water. Channels *x* and *y* must lie in the horizontal plane. If the receiver is lighter than water, buoyancy will cause it to stretch the bungee cord attached to the bottom of the receiver on the z axis, in which case the *z* axis of the receiver will automatically assume a vertical position.

If the receiver is heavier than water, it will sink in water under the effect of differential gravity and stretch the bungee cord attached to its top on the *z* axis. Here too the *z* axis of the receiver also automatically assumes a vertical position. A vector receiver suspended in water on a bungee cord is an oscillatory system with a resonant frequency. The natural frequency of such an oscillating system must lie outside the operating frequency range so as not to affect the receiver's phase response. For example, for a receiver of density $\rho \approx 1.5$ g/cm^3 and 20 cm in diameter, the unbalanced force in water is around 2 kgf. The suspension that the author used consists of two vertical bungee cords both anchored at the top of the vector receiver at an angle of 45° to each other, and an elastic brace between them to achieve a resonant frequency of less than 1–2 Hz (Figs. 2.5 and 2.6). Oscillations of a suspended receiver are extremely difficult to calculate analytically [14], which is why the author would usually suspend vector receivers in a pool of sea water or in a marine environment. When measuring at frequencies below 50 Hz, we need to consider the potential effect of suspension reaction on amplitude and phase response of the vector receiver. In some cases, researchers set the lower end of the operating frequency range at 5–10 times the suspension resonance frequency to make sure measurements are reliable and accurate.

Another problem has to do with diffraction between the closely spaced acoustic pressure receiver and vector receiver. We need to assess how much their scattered fields distort the amplitude and phase of the acoustic field parameters we are interested in. The vector receivers we used were $D = 0.2$ m and hydrophones 0.05 m in diameter. Hence it follows that with such ratios of the two receiver diameters, only diffraction on the rigid sphere of the vector receiver need be taken into account. With the vector receiver $D = 0.2$ m in diameter and operating frequencies up to 1000 Hz, the effect of diffraction on the sphere on field amplitude can be neglected [1, 13]. Often, errors of differential phase characteristics of the p channel and channels x, y, z in the operating frequency range must be limited to less than $\pm$ 3°.

To do this, we build the p channel of six hydrophones in symmetrical arrangement relative to the geometric centre of the vector receiver. If we place the centre of the vector receiver at the origin of a Cartesian coordinate system $O(0,0,0)$, then the six identical hydrophones must be arranged in pairs along the x, y, z axes at equal distances from the centre: ($+x0$, $-x0$); ($+y0$, $-y0$); ($+z0-z_0$). Electrical signals from the six hydrophones arrive at an adder and form a single p channel of the combined receiver, whose directivity pattern is a sphere. Most of the hydrophones were flat, 0.05 m or less in diameter; these were installed no further than their diameter away from the surface of the vector receiver sphere, or 0.03–0.05 m [17, 21].

Combined receiver designs exist in which four hydrophones are mounted directly on the body of the vector receiver. Paper [18] describes a design in which four hydrophones are located at the vertices of an imaginary tetrahedron inscribed in the sphere of the vector receiver to eliminate phase advance and achieve a circular directivity pattern. The authors argue that this arrangement of sound pressure receivers relative to the geometric centre of the vector receiver makes the combined receiver design spherically symmetrical.

The combined receiver described above is placed in an instrument module, whose design depends on whether the combined receiver is intended for a bottom-mounted or a free-drifting measuring system.

2.6 Combined Underwater Acoustic Receiving Systems

2.6.1 Features of Acoustic Measurements in the Ocean

Studying acoustic fields in the real-world underwater environment is methodologically and technologically challenging. This is mostly because pressures and particle velocities in an acoustic wave are much smaller than their ambient hydrodynamic perturbations.

Rather than only changes in pressure or particle velocity of the acoustic field, a receiver in a real-world ocean environment will also record random non-wave interference caused by the oncoming flow of fluid. Pulsations resulting from the flow of fluid around the receiver are known as flow noise, or interference noise. A stationary flow of fluid is free from time-periodic pulsations, but these do develop on the receiver itself due to vortex formation. The spectrum of the flow noise extends from infrasonic to low audible frequencies (up to about 200 Hz). To eliminate the flow noise, the sound receiver must be placed inside a housing—a device transparent to acoustic waves in which the speed of oncoming fluid flow must be zero [17–20].

The housing can distort sensitivity and directivity of the combined receiver [20]. It must be acoustically transparent in the frequency region of interest. Phase distortions that occur when an acoustic wave passes through the housing must be negligible. For this to be true, the housing must be free of large metal parts that can distort the field by sound reflection. The effect of the housing shell on the near field and the impedance of the vector receiver is negligible if the distance between the centre of the vector receiver and the wall of the housing shell is at least $2D$, where D is the diameter of the vector receiver. When these conditions are met, it can be assumed that the combined receiver is in an infinite medium [13].

Let us estimate the various noises that can interfere with the acoustic signal. A simple example shows that even minor vertical displacements of the hydrophone can generate interference noise comparable to the ocean noise. The pressure of low-frequency ocean noise (f > 10 Hz) is on the order of 10^{-3} Pa; for static pressure to change by this much, a hydrophone would need to move vertically a mere 0.01 cm. Apart from pressure fluctuations caused by vertical movements of a hydrophone, interference noise may contain components that are due to the relatively high (about $10^{-3\ \mathrm{V}}\ \mathrm{s}^2\ \mathrm{m}^{-1}$) vibration sensitivity of hydrophones. In pressure terms, this sensitivity is $5^{\ \mathrm{Pa}}\ \mathrm{s}^2\ \mathrm{m}^{-1}$. Vector field characteristics, such as particle velocity (pressure gradient), greatly complicate the measurements.

Here is an example. Let us write out the relationship between the amplitudes of sound pressure p_m, particle displacement ξ_m, particle velocity V_m and particle acceleration a_m for a plane monochromatic wave:

$$p_m = \rho c \omega \xi_m = \rho c V_m = \frac{\rho c}{\omega} a_m, \tag{2.39}$$

where ρc is the acoustic impedance of the medium; ω is the angular frequency of the monochromatic wave.

The spectral density of ambient noise in the 400–1000 Hz frequency range at surface wind speeds of 7.5–9.8 m/s (Fig. 2.10) is near the spectral level to 60 dB re 1 μPa and is equivalent to an acoustic pressure of 10^{-3} Pa, or in acoustic intensity terms $I = 0.667 \times 10^{-15\ \mathrm{W}}/\mathrm{m}^2$.

Assuming $f = 1000$ Hz and $p_m = 10$ Pa, we find:

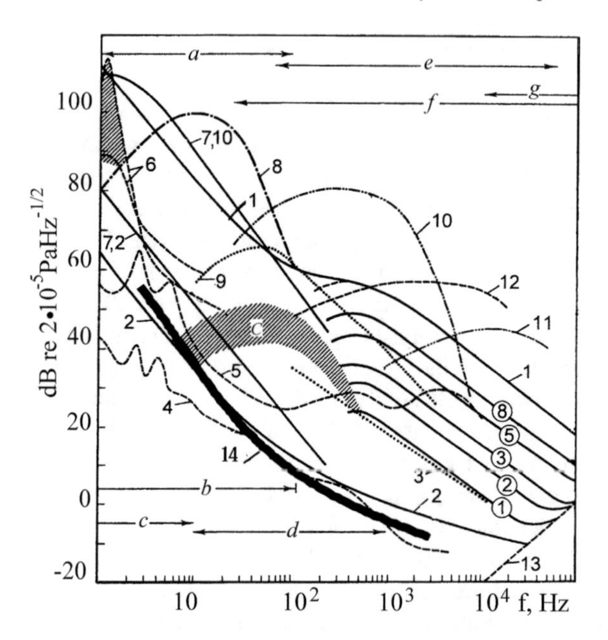

Fig. 2.10 Energy spectra of ambient underwater noise [20] and self-noise spectrum of a combined receiver [17, 21]. 1, 2—maximum and minimum levels of dynamic noise; 3—noise during a calm after Knudsen; numbers in circles are wind speed force (Beaufort), a parameter of the spectra; 4, 5—under-ice noise; 6—seismic background; 7,2, 7,10—interference noise; 8—volcanic eruption spectrum (averaged); 9—navigation noise (area C: shipping lane noise); 10, 11—*Sciaenidae* and shrimp; 12—heavy rain; 13—thermal noise; 14—self-noise of the combined receiver. *a*—frequency region of seismic background, explosions, earthquakes and ice breakup; *b*—turbulent noise; *c*—surface waves; *d*—industrial activity; *e*—cavitation and rain; *f*—thermal noise; *g*—biological noise

$$\xi_m \approx 10^{-10}\,\text{cm},\ V_m \approx 10^{-7}\,\text{cm}/s,\ a_m \approx 10^{-4}\,\text{cm}/s^2. \tag{2.40}$$

It is evident from (2.40) that to measure ambient underwater acoustic noise, we must deal with exceptionally small displacements and particle velocities. Low intensities of ambient noise transform into extremely small displacements in the acoustic wave. At the same time, hydrodynamic movement of the medium (surface waves, internal waves, etc.) and possible movements of structural components of the instrument systems caused by hydrodynamic disturbances can reach several meters. It then follows that the measurement challenge consists in the small absolute values of the acoustic quantities to be measured. This monograph will not discuss the mechanism of flow noise generation or its energy spectrum because they are not the subject of the author's research. Below is a description of the design of bottom-mounted and free-drifting combined systems with self-noise levels (Fig. 2.10, curve 14) capable of operating in the range from calm seas (curve 3) to surface wind speeds of up to ~ 18 m/s (Fig. 2.10, curve 8).

2.6.2 Bottom-Mounted Combined Receiving Systems

In full-scale acoustic experiments, the combined receiver must be positioned at a predetermined observation point in the ocean waveguide. Figures 2.11 and 2.12 illustrate the various types of bottom-mounted combined systems designed by the author. Bottom-mounted systems are physically connected to the ocean floor, and their receiving modules can be located on or near (1.5–3.0 m above) the bottom or in the water column (tens to hundreds of meters above the bottom) supported by buoyancy.

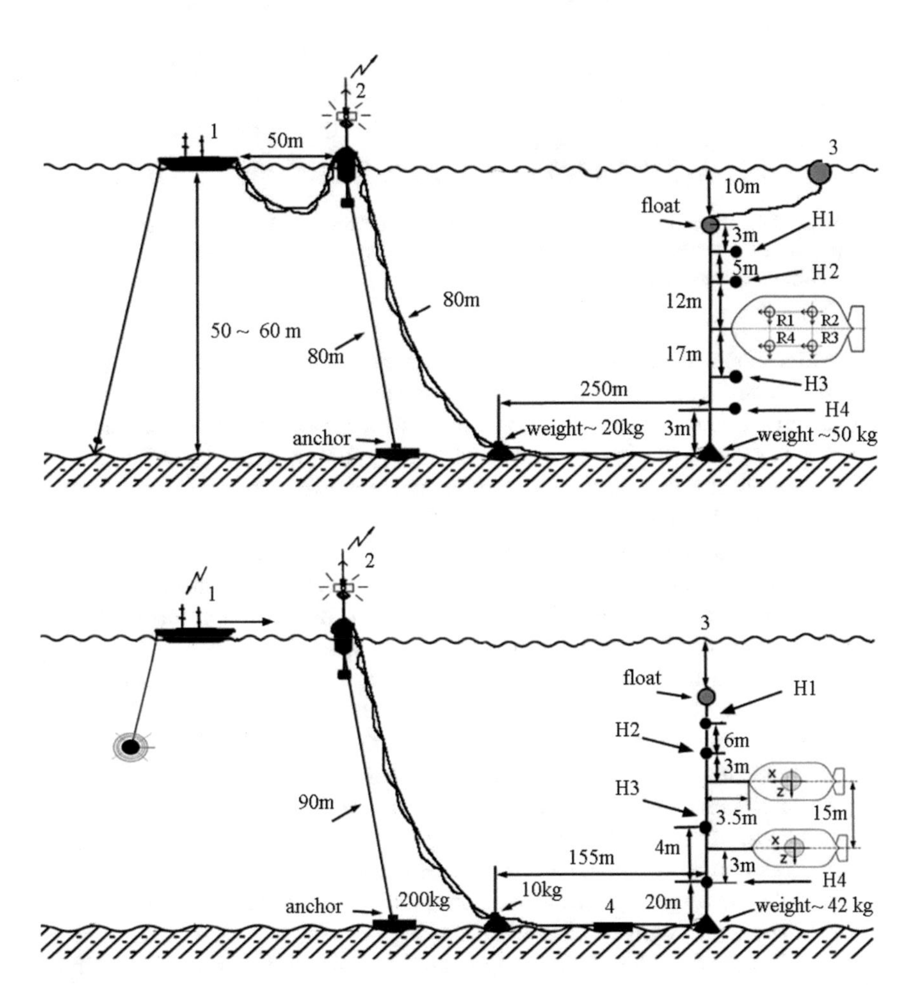

Fig. 2.11 Different types of combined acoustic receiver systems: 1—source and receiver research vessel; 2—RF transmitter buoy; 3—subsurface floats; 4—weight; $H_{1,2,3}$—vertical receiving hydrophone array. The receiving modules are positioned inside the waveguide water column

Fig. 2.12 Eight-channel bottom-mounted combined receiving modules with housings removed. Components of a combined bottom-mounted receiver system. Deployment depth: up to 150 m. Vityaz Bay, Peter the Great Gulf

Bottom-mounted receiving systems were used in coastal waters at depths of 300 m or less. The main difficulty that designers of bottom-mounted systems had to overcome was the potentially significant flow noise caused by bottom and tidal currents. In the coastal zone, there is usually heavy fishing and cargo traffic; this too greatly complicates the study of underwater ambient noise and signal near or below underwater ambient noise.

A bottom-mounted system resting on the seafloor is a metal frame in the form of a tripod to which a combined receiver is attached. Bottom currents cause the metal structure of the bottom station to vibrate and turbulent currents to develop around it, causing vibrations and low-frequency noise. Vibrations of the bottom station's components in the incident flow of water, together with flow noise, reach the combined receiver and interfere with sound reception. To isolate the combined receiver from this source of interference, the bottom station is designed to have these components: an external housing, an internal housing and a dual-link receiver suspension system. Inside the outer housing, the fluid flow velocity is zero. The significant added mass of water inside the housing eliminates body vibrations, and the tripod structure is covered with soft fleece fabric.

The outer housing is a cylinder or an axially symmetric ellipsoid displacing 1–3 m^3. The metal frame of the outer housing is covered with soft fleece fabric or fine nylon mesh. Other designs are possible, in which the external housing is suspended inside a tripod [22].

The main purpose of the external housing: completely eliminate the bottom current (inside the housing, fluid velocity must be zero); significantly reduce vibrations transmitted from the station housing to the combined receiver; if the bottom is rough, the axis of the tripod may tilt; if that happens, gravity will right the major axis of the external housing. The bottom station is so designed that when the tripod's axis tilts up to 30° angle from the plumb line, the axis of the external housing will assume a vertical position.

The inner housing is a sphere or an ellipsoid. Its frame is also covered with soft fleece fabric. The inner housing suspended at its uppermost point is a pendulum whose natural frequency in water doesn't exceed 0.1 Hz. The combined receiver is suspended inside the housing. The frame of the inner housing is made of positively buoyant syntactic foam. The inner housing together with the combined receiver must either have a very nearly neutral buoyancy or be slightly positively or slightly negatively buoyant. This almost completely isolates the internal housing from oscillations and vibrations and reduces interference by 10–20 dB across several octaves. We should note that hydrophone-based receiver systems need not be so complex. The volume of the module's external housing has a crucial role to play. The significant added mass of the water is what stabilises it in the water column.

2.6.3 *Free-Drifting Combined Telemetry Systems*

Free-drifting autonomous telemetry receiving systems developed by the author are coupled to the surface of the ocean. Their joint movement with the surrounding water parcels reduces flow noise to a point where it no longer interferes with sound reception. There is another method of free-drifting combined reception: autonomous neutrally buoyant floats [10]. The main source of interference with sound reception for receiving systems coupled with the ocean surface is surface waves, which (especially at wind speeds over 10 m/s) exerts significant forces on the cable line, causing it to vibrate and pull. The problem of suppressing the mechanical effect of surface waves on surface-coupled systems is a subject of intense discussion in scientific literature (see, for example, [20]).

The free-drifting combined measuring system consists of an RF transmitter buoy floating on the ocean surface and a long submarine cable line (up to 1500 m long), to which the combined receiving modules are mechanically and electrically connected. Forced displacements of the receiving module as a result of poor decoupling from surface waves and water flow about the receiving module are a source of non-acoustic interference. Noise of the water flowing over the cable and the receivers or turbulence noises originating from outside the receiver can far exceed the real low-frequency noise.

Flow noise can be significantly reduced if the receiving system moves with the flow of the surrounding fluid—i.e. if the relative velocity of the fluid flow over the receiving module is zero. Given the above, the movement of the module relative to the surrounding water should be minimised, and in the extreme case the receiving module should be 'frozen' into the water and move with it. This idea led us to design a free-drifting autonomous telemetry combined receiving system. Because the receiving system is coupled to the surface of the ocean, it must be designed to suppress underwater oscillations and vibrations of the system caused by the movement of sea surface energy.

Figure 2.13 is a schematic of a free-drifting eight-channel receiving system with two four-channel combined receivers. A watertight container (1) and a radio transmitter (2) form the surface component of the receiving system, called the RF transmitter buoy. It weighs no more than 50 kg and causes little drag.

To dampen the twitching of the vertical line CB by the RF transmitter buoy bobbing on the waves (Fig. 2.13a), sections of the horizontal cable line CK were made to sag into the water using a series of unevenly spaced floats (3).

Cable line CK is attached at point K located at the level of the container's metacentre (1). This way, rotational oscillations of the buoy don't tug on the cable. Vertical submarine cable line BC is balanced as follows. Underwater, it carries a cylindrical subsurface float (4) made of syntactic foam weighing ~ 100 kg. Subsurface float

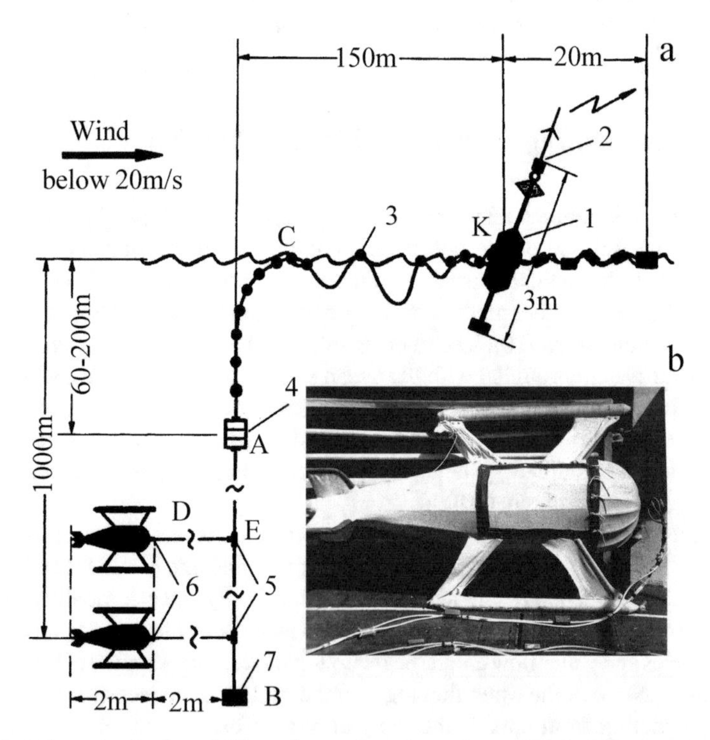

Fig. 2.13 Schematic of a free-drifting autonomous telemetric combined eight-channel receiving system **a** and a four-channel combined module **b**. Legend: **a** 1—watertight container with electronics and a power source; 2—radio transmitter; 3—horizontal cable line with floats; 4—deepwater subsurface float; 5—cable terminal boxes; 6—combined receiving modules; 7—weight. **b** Four-channel neutrally buoyant combined module ready to launch [5, 17, 21]

(4) acts as an inertia dampener. At point A, the negative buoyancy of line AB is 10–15 kgf. This negative buoyancy is compensated by positive buoyancy of a long string of small floats of varying buoyancy strung on line AC. The floats have progressively lower buoyancy (volume) from point A to point C. The upper float at point C (usually painted bright red) has a buoyancy of 0.2 kgf. If vertical cable line CB is properly balanced, the last bright red float must surface in calm seas. In rough seas, this float sinks when a wave crest approaches and resurfaces in troughs between wave crests. Because the buoyancy of floats on line AC near the surface is small and the inertial mass of line CB is significant, surface waves have little effect (in the form of vertical jolts and longitudinal oscillations) on line AB at point A. The large added mass of subsurface float (4) creates neutral buoyancy at the EB section. Experiment has shown that in this design, the vertical channel z, the one most sensitive to vertical movements, will be completely protected from vertical interference noise if positioned at point B.

Apart from vertical tugs, line CB may experience transverse oscillations. To suppress transverse oscillations, the entire line CB is wrapped in evacuated foam sheathed in plastic mesh. The foam cylinder is ~ 0.1 m in diameter in air. Also, thin ribbons of soft fabric ~ 1 m long are attached to line CB at random points.

As a result, the entire part of the system below point A has a significant inertial mass to keep it 'stationary' relative to the surrounding fluid and prevent significant oscillations in the long line EB. However, if a combined receiver is located at point B in this system, then at surface wind speeds of > 12 m/s transverse and longitudinal oscillations of line BE increase to a point where ambient noise levels no longer register at frequencies below 200 Hz. To more completely decouple them from oscillations and vibrations of the vertical line EB, receiving modules (6) are connected to the vertical cable line by a 10–25 m-long neutrally buoyant horizontal cable DE. Receiving modules (6) too are neutrally buoyant. The instrument module is an axisymmetric neutrally buoyant body with a combined receiver inside. Also housed inside the module are roll, pitch and depth sensors and preamplifiers (Fig. 2.13b).

The combined receiving acoustic module for measurements in the 1–1000 Hz frequency range is designed as follows. The module consists of a structure and a housing shell. The teardrop-shaped rigid structure of the housing is covered with soft fleece fabric. The module and the horizontal section of the cable are neutrally buoyant.

As full-scale experiments have shown, in free drift the water current speed relative to line CB doesn't exceed 0.02–0.03 m/s, which is enough for the current to orient receiving modules (6) in a steady horizontal position.

Cable line AB can be made up of cable sections of different lengths linked together with sealed connectors. This allows the receiving modules (6) to be placed at different depths depending on the mission. Measurement depths varied from as shallow as 20 m to a maximum of 1000 m [21].

Figures 2.14 and 2.15 show the configuration of the combined receiving module and an actual unit. The module weighs no more than 20 kg in air.

Before each experiment, the module was adjusted for roll, pitch and neutral buoyancy in a purpose-built seawater pool aboard the research vessel. Roll and pitch of the

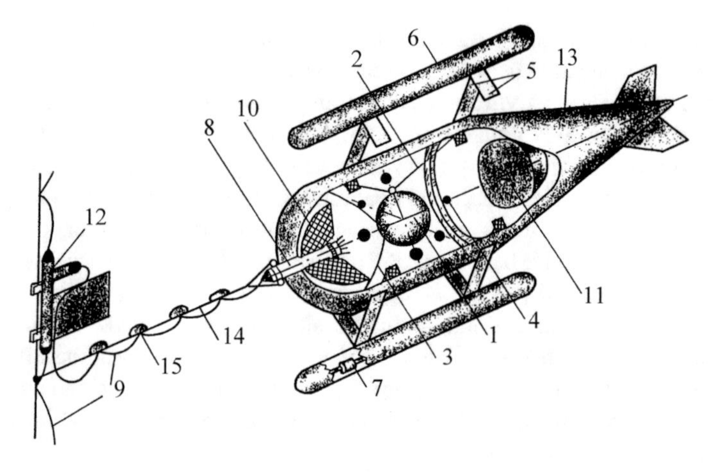

Fig. 2.14 Configuration of the combined acoustic receiving module. Legend: 1—three-component particle velocity sensor; 2—bungee cords; 3—suspension limiter; 4—hydrophones; 5—positive buoyancy subsurface float attachments; 6—positive buoyancy subsurface float; 7—pitch controller; 8—preamplifier container; 9—cable; 10, 11—subsurface floats; 12—cable terminal box; 13—housing; 14—polypropylene line; 15—small floats made of syntactic foam

Fig. 2.15 Combined four-channel receiving modules with housings removed aboard an R/V before being launched

module were set to within ± 5°. These tilts are fully compensated by the combined receiver's freedom of pendulum motion inside the housing.

Once launched in the ocean, a freely drifting system takes 45–60 min to deploy regardless of wind speed and sea state and remains operational up to surface wind speeds of ~ 18 m/s. We should note that although this system is long, it doesn't require a specially equipped vessel to launch and retrieve. The vessel need to only have a ramp and a stern winch to launch and retrieve the system.

2.6.4 Features of Vector Receiver Suspension in Free-Drifting Receiving Systems

An important part of the module's design is the way the combined receiver is suspended inside the instrument module. A vector receiver's operating frequency range is sandwiched between its two resonances: the low frequency and the high frequency. The low-frequency resonance depends on the vector receiver suspension design.

A co-oscillating receiver is held inside the housing on elastic braces and orients in space in a certain way. Because at $\rho \approx 1500$ kg/m^3 the receiver is denser than water, the bungee cords experience the static load of the receiver's differential weight. The high-frequency resonance is due to the design of the piezoceramic transducers of the receiver and lies in the 1.5–2.5 kHz region for the receivers we use. We cap out operating frequencies at 1000 Hz.

With the acoustic pressure receiver and the vector receiver close to each other, we need to assess how much their scattered fields distort the amplitude and phase of the acoustic field parameters of interest. The vector receivers we use are $D = 0.2$ m and hydrophones 0.05 m in diameter. Hence it follows that with such ratios of the two receiver diameters, only diffraction on the rigid sphere of the vector receiver need to be taken into account. With the vector receiver $D = 0.2$ m in diameter and operating frequencies up to 1000 Hz, the effect of diffraction on the sphere on field amplitude can be neglected [1, 13]. The differential phase error of the p channel and channels x, y, z in the operating frequency range can be limited to ± 3° or less if the p channel is made up of six hydrophones symmetrically arranged about the geometric centre of the vector receiver to form a combined receiver. Most of our hydrophones were flat, 0.05 m or less in diameter; these were installed no further than their diameter away from the surface of the vector receiver sphere, or 0.03–0.05 m.

There are several ways to suspend vector receivers [14]. We use our suspension design, proven in numerous full-scale experiments. The suspension is designed as follows. The vector receiver is suspended along the x, y, z axes on six thin nylon filaments that serve to limit the movement of the receiver in the module. The nylon filaments keep the receiver from tilting more than 30° from the vertical. The orthogonal axes x, y, z of the receiver are oriented as follows: the x axis is horizontal and runs along the module's longitudinal axis of symmetry; the y axis is also horizontal;

and the z axis is vertical and points from the surface down to the ocean floor. The bungee cords are all attached to the top of the receiver sphere on the z axis. The other ends of the bungee cords are anchored to the housing shell. When the module is submerged, uncompensated gravity force causes it to assume a position in which the z axis points vertically downwards. The receiver floats up vertically and is held in space only by the vertical bungee cord. To minimise rotation of the receiver about the z axis, the vertical elastic suspension consists of a V-shaped bungee cord with a 90° aperture of the V. The natural frequency of this suspension is less than 1 Hz. In some cases, researchers set the lower end of the operating frequency range at 5–10 times the suspension resonance frequency to make sure measurements are reliable and accurate.

We should mention another way to suppress vibrations. If the sound pressure sensors are mounted on the body of the pressure gradient vector receiver, vibration interference can be processed out. In this case, vibration interference is suppressed by the 90° phase shift of the pressure gradient components relative to the acoustic pressure.

2.6.5 *Vector Receiver Systems on Unmanned Underwater Vehicles (Gliders)*

An underwater glider is a small autonomous unmanned vehicle capable of remaining in the water column for extended periods of time while covering considerable distances. The glider's operating principle is that its dive depth changes as it moves. Nearly silent and compact, it can carry a payload, including a combined acoustic receiving system. The level and anisotropy of underwater ambient noise can be investigated on a global scale in terms of distance and depth using a glider. A glider can be used to continuously monitor the acoustic situation in protected waters. The high noise immunity of vector-phase reception means that a glider can serve as a platform for a system capable of detecting weak signals in both diffuse acoustic field and coherent noise field.

Paper [23] plots the course taken by the glider on missions to detect weak signals and investigate underwater ambient noise (Fig. 2.16).

The research was a part of the project titled 'Systems approaches and innovative technology in shipbuilding; development of new types of marine equipment' (2009–2011) done in partnership with the acoustic centre of the Physics Department of the G. I. Nevelskoi Maritime State University [23–27].

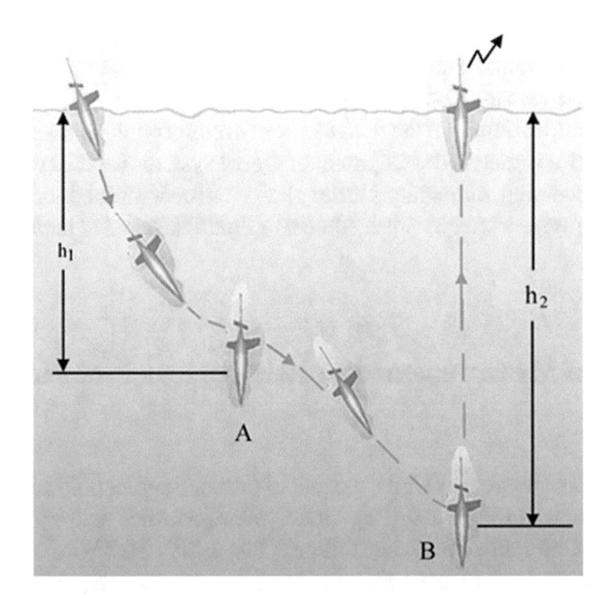

Fig. 2.16 Carrier movement during acoustic measurements. A, B—position of the carrier during measurement

2.7 Counterparts Outside Russia

Particle velocity instruments in underwater acoustics were first mentioned in the 1930s (Germany, Konrad Tamp and Erwin Mayer). The first modern type of co-oscillating hydroacoustic particle velocity sensor was described in 1942. It was manufactured by Bell Telephone Laboratories for the U.S. Navy's hydroacoustic measurement laboratory [28]. English-language terminology in this branch of acoustics was very peculiar. The particle velocity sensor was called a particle velocity hydrophone, and the technique the method of acoustic intensity measurement. English-language terminology has since switched to the nomenclature proposed by Rzhevkin [2, 6]: the technique is called the vector-phase method, and the receivers vector and combined receivers. The vector-phase method has evolved and gained traction abroad, especially in the United States, for both ocean acoustics research and military applications. DIFAR (Directional Frequency Analysis and Recording) is a sonobuoy widely used for submarine detection. The DIFAR sonobuoy is not designed for scientific research. It has a combined receiver consisting of a pressure sensor and two horizontal components of particle velocity. The characteristics of this buoy and its potential uses are described in [29]. Marine Physical Laboratory, Scripps Institution of Oceanography (San Diego, USA), has conducted studies with free-drifting neutrally buoyant floats in the 0.6–20 Hz frequency range [10].

Paper [30] describes combined receivers MOD-1 and VHS-100 and sensors used in the receiving systems. The receiver design and application methodology exactly replicate Russian vector-phase systems [5, 15, 16, 19].

A surge in publications that the field has seen in connection with new piezoelectric materials, laser transducers, development of digital systems for data recording, transmission and processing, unmanned carrier platforms for combined acoustic passive or active receiving systems marks this method as a significant area of modern underwater acoustics.

2.8 Units of Measurement and Relative Levels of Measured Values

In the International System of Units, the unit of pressure p generally accepted in absolute measurements is the pascal (Pa). Underwater acoustics also uses the microbar (μbar). Unit of pressure Pa = newton/m^2 = kg/(m s^2), Pa = N/m^2. One microbar of pressure equals 10^{-6} of the standard atmosphere, μbar = dyne/cm^2 = g/(cm s^2), 1 bar = 1.013 atmosphere. Table 2.1 converts between commonly used units of measurement.

The reference pressure of the decibel scale is the micropascal 1 μPa = 10^{-6} Pa, or 2×10^{-4} μbar equal to 2×10^{-5} Pa. Pressure levels relative to 1 μPa and 2×10^{-4} μbar differ by 26 dB.

Sensitivity of transducers is usually measured in μV/Pa or in dB re 1 V/μPa.

The reference acoustic intensity is defined as the intensity of a flat wave $I = \langle p^2\rangle/\rho c$, where $\langle p^2\rangle$ is the RMS pressure value and ρc is the acoustic impedance of the medium. In water, $\rho c = 1.5 \times 10^5$ g/(cm^{2}s) = 1.5×10^6 kg/(m^{2}s).

Table 2.1 Conversion of pressure values given in pascals into other units commonly used

1 N/m^2 = Pa	μPa	dB Re 1μPa	dB Re 20 μPa	dyne/cm^2 = μbar	atm ≈ bar
10^5	10^{11}	220	194	10^6	1
10^4	10^{10}	200	174	10^5	10^{-1}
10^3	10^9	180	154	10^4	10^{-2}
10^2	10^8	160	134	10^3	10^{-3}
10^1	10^7	140	114	10^2	10^{-4}
1	10^6	120	94	10^1	10^{-5}
10^{-1}	10^5	100	74	1	10^{-6}
10^{-2}	10^4	80	54	10^{-1}	10^{-7}
10^{-3}	10^3	60	34	10^{-2}	10^{-8}
10^{-4}	10^2	40	14	10^{-3}	10^{-9}
10^{-5}	10^1	20	− 6	10^{-4}	10^{-10}
10^{-6}	1	0	− 26	10^{-5}	10^{-11}

The reference intensity level at 1 μPa pressure is $I = 0.667 \times 10^{-18}$ W/m^2. The reference intensity level for 1 μbar = 0.1 Pa equals $I_0 = 0.667 \times 10^{12}$ W/cm^2 = 0.667×10^{-8} W/m^2.

2.9 Conclusions

The vector-phase method is clearly necessary and sufficient in ocean acoustics. The author's combined vector-phase receiving systems for shallow seas and deep open ocean have evolved since 1980 through full-scale acoustic experiments in real ocean environments. The research has laid the groundwork for an advanced new branch of underwater acoustics called vector acoustics of the ocean. The frequency range of studies was 1–1000 Hz; measurement depths up to 1000 m; sea state up to 6 on the Beaufort scale at wind speeds up to ~ 18 m/s.

Receiving systems developed by the author are original designs; algorithms have been developed for recording multichannel digital data (up to 32 channels). Technical capabilities of receiving systems—sensitivity and directivity of vector channels, sensitivity relation coefficient, dynamic range of receiving circuits, electronic self-noise—have allowed underwater ambient noise and signal to be studied over a wide range of frequencies.

References

1. S.N. Rzhevkin, On vibrations of bodies immersed in fluid and acted on by a sound wave. MSU Vestnik **1**, 52–61 (1971). (in Russian)
2. L.N. Zakharov, S.N. Rzhevkin, Vector-phase measurements in acoustic fields. Akusticheskij Zhurnal (Acoust. Phys.) **20**(3), 393–401 (1974). (in Russian)
3. A.A. Kharkevich, *Theory of Transducers* (Gosenergoizdat, Moscow, 1948). (in Russian)
4. C. Leslie, J. Kendall, J. Jones, Hydrophone for measuring particle velocity. JASA **28**, 711 (1956).
5. V.A. Shchurov, *Vector Acoustics of the Ocean* (Vladivostok, Dalnauka, 2003). (in Russian)
6. S.O. Vorobyov, V.I. Sizov, Vector-phase structure and vector-phase method of describing and analysing random acoustic fields. Akusticheskij Zhurnal (Acoust. Phys.) **38**(4), 654–659 (1992). (in Russian)
7. L.D. Landau, E.M. Lifshitz, *Fluid Mechanics* (Nauka, Moscow, 1986). (in Russian)
8. S.N. Rzhevkin, *A Course in the Theory of Sound* (MSU, Moscow, 1960). (in Russian)
9. P.M. Morse. *Vibration and Sound* (McGraw-Gill Book Company, N.Y., 1948)
10. G.L. D'Spain, W.S. Hodgkiss, G.L. Edmonds, The simultaneous measurement of infrasonic acoustic particle velocity and acoustic pressure in the ocean by freely drifting Swallow floats. IEEE J. Oceanic. Eng. **16**(2), 195–207 (1991)
11. L.N. Zakharov, V.I. Ilyichev, S.A. Ilyin, V.A. Shchurov, Vector-phase measurements in ocean acoustics, in *Problems of Ocean Acoustics*, eds. by L.M. Brekhovskikh, I.B. Andreeva (Nauka, Moscow, 1984), pp. 192–204. (in Russian)
12. V.I. Ilyichev, V.A. Shchurov, V.P. Dzyuba, Investigation of the acoustic noise field of the ocean by vector-phase methods, in *Acoustics of the Ocean Environment* (Nauka, Moscow, 1989), pp. 144–152. (in Russian)

13. S.N. Rzhevkin, Near field and impedance of a sphere oscillating near a hard and a soft partition. Akusticheskij Zhurnal (Acoust. Phys.) **24**(1), 143–146 (1978). (in Russian)
14. V.E. Ivanov, Analysis of the effect of suspension of a vector receiver on its characteristics. Akusticheskij Zhurnal (Acoust. Phys.) **34**(1), 95–101 (1988). (in Russian)
15. V.A. Shchurov et al., A device for measuring the parameters of noise sources, USSR invention certificate 953468, Bull. 31, 1982. (in Russian)
16. V.A. Shchurov et al., Invention certificate 321205.1990 (claimed since 10.04.1989)
17. V.A. Shchurov, *Research Report of the 9th Voyage of R/V Akademik Lavrentyev* (Archives of the Pacific Oceanological Institute, Vladivostok, 1987)
18. A.N. Zhukov, A.N. Ivannikov, V.V. Isaev, V.N. Nyunin, O.S. Tonakanov, A.V. Shiryaev, A sensor for acoustic measurements, in *10th National Acoustics Conference*, July-1–5 (Moscow, 1983), pp. 59–61. (in Russian)
19. G.K. Skrebnev, *Combined Hydroacoustic Receivers* (Elmor, Saint Petersburg, 1997). (in Russian)
20. A.V. Furduev, *Oceanic Noise*, in *Acoustics of the Ocean,* eds. by L.M. Brekhovskikh (Nauka, Moscow, 1974), pp. 615–692. (in Russian)
21. V.A. Shchurov, *Research Report of the 14th Voyage of R/V Akademik Vinogradov* (Archives of the Pacific Oceanological Institute, Vladivostok, 1989)
22. L.N. Zakharov, V.A. Shchurov, Study of ocean noise by vector-phase methods. Metrology-DVO research project report. Vol. 1 (Pacific Oceonological Institute, Far Eastern Scientific Centre, USSR Academy of Sciences. Vladivostok, 1980). (in Russian)
23. V.A. Shchurov, E.N. Ivanov, S.G. Shcheglov, A.V. Cherkasov, An underwater glider for monitoring acoustic vector fields (in Russian): pat. 106880 U1 RF, No. 2011108806; app. 09.03.11; publ. 27.07.11, Bull. no. 21 (utility model)
24. S.G. Shcheglov, Underwater glider (versions): pat. 122970 U1 RF, no. 2012118807; appl. 04.05.12; publ. 20.12.12, Bull. 35 (utility model)
25. S.G. Shcheglov, A.S. Lyashkov, Underwater glider (versions): pat. 124245 U1 RF, no. 20121186660, appl. 04.05.2012; publ. 20.01.2013, Bull. 2 (utility model)
26. S.G. Shcheglov, Underwater glider (versions): pat. 2490164 C1 RF, no. 2012118812; appl. 04.05.2012; publ. 20.08.2013, Bull. 23 (invention)
27. S.G. Shcheglov, Underwater glider (versions): pat. 176835 U1 RF, no. 2017119727; appl. 05.06.2017; publ. 31.01.2018, Bull. 4 (utility model)
28. L. Bobber, *Hydroacoustic Measurements* (Mir, Moscow, 1974). (in Russian)
29. G.L. D'Spain et al., Vector sensors and vector sensor line arrays: comments on optimal array gain and detection. J. Acoust. Soc. Am. **120**(1), 171–185 (2006)
30. D.R. DallOsto, Properties of the Acoustic Vector Field in Underwater Waveguides. A dissertation for the degree of Doctor of Philosophy, 2013

Chapter 3
Phenomenon of Compensation of Intensities of Reciprocal Energy Fluxes

3.1 Introduction

In the real ocean, acoustic energy from an isolated sound source takes several different paths to reach the observation point. If monochromatic waves from the same source propagate by several different paths, their superimposed energies add coherently, causing energy flux density to oscillate at a certain frequency in terms of level as well as, crucially, direction.

If the intersecting wave processes are statistically independent, such as underwater ambient noise and interfering tonal or broadband signal, the interaction of their intersecting energies can be observed by measuring the net vector intensity of this process. As is known, the result of superposition of intersecting waves depends on their orientation relative to each other. Interaction of superimposed travelling plane waves can be sufficiently well described by acoustic pressure [1]. For two plane waves travelling in the same direction, the total energy flux density is not additive. Indeed, the energy flux density of the sum of two waves $p = p_1(t-x/c)+p_2(t-x/c)$ is

$$I = \frac{1}{\rho c}p^2 = I_1 + I_2 + \frac{2}{\rho c}p_1 p_2 \tag{3.1}$$

where $I_1 = p_1^2/\rho c$; $I_2 = p_2^2/\rho c$; p_1 and p_2 are amplitudes of the first and the second wave; ρ is medium density; and c is sound speed. Additivity holds for energies of monochromatic waves of different frequencies averaged over a long period of time or for statistically independent travelling waves.

Energy flux density in the region where two plane waves travelling in opposite directions intersect always equals the difference of their energy flux densities. For example, the sum of waves travelling in the $+\,x$ and in the $-\,x$ directions, $p_1 = p_1(t-x/c)$ and $p_2(t+x/c)$, is $p = p_1 + p_2$, $v = v_1 + v_2 = \frac{1}{\rho c}(p_1 - p_2)$, hence

V. A. Shchurov, *Movement of Acoustic Energy in the Ocean*,
https://doi.org/10.1007/978-981-19-1300-6_3

$$I = pv = (p_1 + p_2)\frac{1}{\rho c}(p_1 - p_2) = I_1 - I_2. \tag{3.2}$$

It follows from (3.2) that if $p_1 = p_2$, then $I_1 - I_2 = 0$. This means that an anomaly unrelated to interference may occur in an area of the field if $p = p_1 + p_2 \neq 0$, $I = I_1 - I_2 = 0$. A standing wave is a classic example of superposition of two coherent waves travelling in opposite directions. The standing wave results from interference, but $I_1 - I_2 = 0$ can be met for incoherent waves, for waves of different frequencies, and for noise and signal. We have termed this phenomenon 'compensation of reciprocal energy fluxes' [2–4].

This chapter goes through real-ocean examples of energy movement in the region of intersecting acoustic waves using the vector-phase method.

Suppose two locally plane waves arrive an observation point from directions $\boldsymbol{s}_1$ and $\boldsymbol{s}_2$, the net energy flux along a direction $\boldsymbol{s}_0$ can be written as:

$$I_{s_0} = \frac{1}{2}p_1 v_1 \cos\theta_1 + \frac{1}{2}p_2 v_2 \cos\theta_2, \tag{3.3}$$

where $I_1 = \frac{1}{2}p_1 v_1$ and $I_2 = \frac{1}{2}p_2 v_2$ are average energy flux densities of the first and the second locally plane waves, respectively; θ_1 and θ_2 are angles between direction s_0 and directions $\boldsymbol{s}_1$ and $\boldsymbol{s}_2$, respectively.

Consider compensation along the z axis. Let $\boldsymbol{s}_0$ point in the $+\,z$ direction. Dynamic noise energy flux intensity is $\frac{1}{2}(p_1 v_{1,+z})_N$ and $\theta_1 = 0°$, $\cos\theta_1 = 1$. For the oncoming flow of signal energy reflected from the bottom, intensity equals $-\frac{1}{2}(p_2 v_{2,+z})_S$, and $\theta_2 = 180°$, $\cos\theta_2 = -1$. Therefore, the net averaged energy flux along the z axis is:

$$I_z = \frac{1}{2}(p_1 v_{1,+z})_N - \frac{1}{2}(p_2 v_{2,+z})_S = I_{+z,N} - I_{-z,S}. \tag{3.4}$$

Similarly, for two energy fluxes propagating in opposite directions along the $+\,x$ axis (horizontal energy flux of dynamic noise) and the $-\,x$ axis (energy flux from a local source):

$$I_x = \frac{1}{2}(p_1 v_{1,+x})_N - \frac{1}{2}(p_2 v_{2,+x})_S = I_{+x,N} - I_{-x,S}. \tag{3.5}$$

Average net components I_z and I_x of energy flux density (3.4) and (3.5) can vanish if their terms cancel each other out. If they do, full compensation of two reciprocal fluxes of acoustic energy will result.

If the averages of the terms are not exactly equal but are of the same order of magnitude, partial (incomplete) intensity compensation will be observed. In summary, intensities of reciprocal energy fluxes can cancel each other out in coherent reciprocal flows of energy as well as in statistically independent random wavefields.

3.2 Experimental Observations of Intensity Compensation

First indications of intensity compensation the author observed during the 1978 Kuril–Kamchatka research cruise over a wide range of dynamic underwater noise at frequencies emitted by R/V *Callisto*. In his final report, the author couldn't explain these anomalies, which presented as 'dips' in cross-spectra S_{pV_x}, S_{pV_y} and S_{pV_z} [5]. In the deep open ocean and in shallow water, this phenomenon was observed in intersecting acoustic fields from various sources: between signals of different frequencies, between signal and underwater ambient noise, etc. Over the years, experiments were conducted in expeditions to central and northwestern Pacific, the South China and the Philippine Seas. The experiments studied a range of depths, frequencies and radiation levels in different hydrological settings. The vessel made passes relative to a combined receiving telemetry system with receiving modules set at various depths. In field experiments, we observed the phenomenon we termed 'compensation of reciprocal energy fluxes' in equal or nearly equal reciprocal flows of acoustic energy, whether coherent or not.

3.2.1 Design of Experiment in the Deep Open Ocean

Let us consider an experimental design in the deep open ocean that brings out the phenomenon of compensation of intensities of reciprocal energy fluxes. The experiment described below was conducted on 18 May 1989 in central South China Sea aboard the R/V *Akademik Vinogradov*. As mentioned earlier, 'dips' in cross-spectra had been observed since 1978, but the experiment in question was organised so that the entire compensation process was controlled for.

Of the most interest is the case of compensation of energy of ambient dynamic noise by the reciprocal flow of energy from a local tonal or noise-like source. Underwater ambient dynamic noises are known to have vertical as well as horizontal energy flows [4]. In the ocean waveguide, these energy flows may face competition from energy flowing from man-made sources, such as ships, sound projectors, etc., which may result in compensation of intensity of interacting energy flows.

Consider compensation of vertical energy flow of dynamic surface noise by the opposite flow of energy from a local underwater tonal source. Coordinate axes of the combined receiver are oriented as follows: z axis is vertical and points from the surface to the bottom; x and y axes are horizontal. The x axis points in the direction of surface wind. The projector is 3.5 km away from the receiving system at a depth of 60 m. The support vessel is adrift and silent. A tone is emitted at $f_0 = 402$ Hz. The level of the tone signal is adjusted in real time based on feedback coming from the telemetry receiving system to the laboratory. Radiated power is adjusted to cancel out the average noise power flux at the given frequency $(I_{+z,N})_{f_0}$ with the average signal power flux $(I_{-z,S})_{f_0}$ at the observation point. As a result, the compensation phenomenon was observed in a full-scale experiment in real time. The experiment also used the noise of a vessel moving past the receiving system. The measuring system consisted of two combined receivers located at depths of 250 and 500 m. Experimental conditions: water depth 3600 m; axis of the SOFAR channel at 1200 m; surface sound speed greater than bottom sound speed; wind speed 12 m/s; steady-state surface waves, swell and wind oriented in the same direction.

3.2.2 Example of Vertical Compensation of Tone Signal and Underwater Ambient Noise Along the Z Axis

The combined receiving system is in the near field of the projector. Figure 3.1 illustrates compensation of z components of intensities between reciprocal energy fluxes

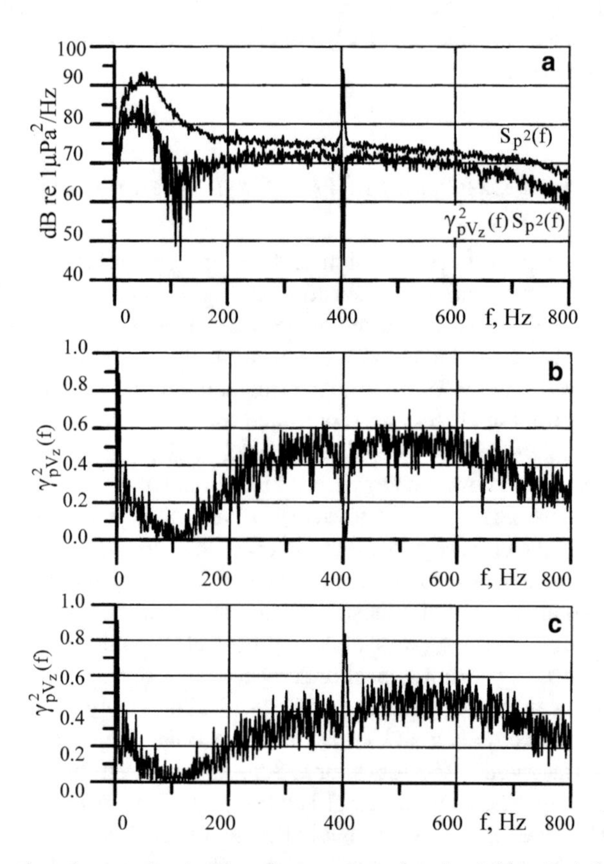

Fig. 3.1 Interaction of a tone signal with underwater acoustic noise field. **a** $S_{p^2}(f)$—power spectrum of noise pressure; $S_{pV_z}(f) - z$ component of the cross-spectrum of acoustic noise; **b** and **c** $\gamma^2_{pV_z}(f)$—frequency coherence function. Tone frequency: 402 Hz. Measurement depth: 500 m. Analysis bandwidth: 1 Hz, averaging time: 60 s, exponential averaging

of dynamic surface component of ambient noise and a tone signal at $f_0 = 402$ Hz reflected from the bottom.

Figure 3.1a shows spectra $S_{p^2}(f)$ and $S_{pV_z}(f)$. The units of spectra $S_{pV_z}(f)$ and $S_{p^2}(f)$ are the same because $S_{pV_z}(f) = \gamma^2_{pV_z}(f)S_{p^2}(f)$ (Chap. 1). In the power spectrum of acoustic pressure $S_{p^2}(f)$, signal exceeds noise by ~ 20 dB at $f_0 = 402$ Hz. But the spectrum of the z component of power has a ~ 28 dB 'dip' at this frequency. At the time of the experiment, the author already knew what result to expect. In previous experiments, especially if 'dips' occurred simultaneously at different frequencies, it was assumed that the receiver system had failed. The movement of energy in this experiment is as follows. Dynamic noise energy propagates from the surface to the bottom in the $+ z$ direction, and its intensity has a z component of $I_{+z,N}$.

The coherence function $\gamma^2_{pV_z}(f)$ of ambient noise rises to 0.5 in the frequency range of 300–600 Hz, meaning that the surface noise is partially coherent (Fig. 3.1b, c). The energy flux of the local source (projector) reflected from the bottom has a z component of intensity of $-I_{-z,S}$. During the experiment, projector power output was varied to equalise the energy flux density $I_{-z,S}$ of the signal at $f_0 = 402$ Hz with that of the noise $I_{+z,N}$ at the given frequency f_0. With $I_{+z,N}$ and $I_{-z,S}$ equal, we have $I_{+z,N} - I_{-z,S} = 0$. In this case, a ~ 28 dB 'dip' occurs in the cross-spectrum $S_{pV_z}(f)$ at $f_0 = 402$ Hz. At the compensation frequency, $\gamma^2_{pV_z}(f_0) = 0$, meaning that the coherent component of ambient noise is completely suppressed by the local source's field (Fig. 3.1b). Increasing the power of the signal reflected from the bottom makes $I_{-z,S} > I_{+z,N}$, breaking the compensation condition $I_{+z,N} - I_{-z,S} = 0$ (Fig. 3.1b). In this case, phase difference $\Delta\varphi_{pV_z}(f)$ at the frequency f_0 differs from that in the noise by 180° [4].

Note that on the 'sidelines' of the tone signal, where signal strength doesn't reach the required level, some of the 'dip' remains. This result opens up exciting possibilities in the practical application of this phenomenon to observe (detect) weak signal in underwater ambient noise, in partially coherent interference noise, etc. [2–4].

3.2.3 Example of Horizontal Compensation in the Shallow Water Waveguide

3.2.3.1 Experimental Conditions

Let us now consider the experiment in intensity compensation of reciprocal energy fluxes of tonal signals at 86, 123 and 163 Hz and partially coherent broadband interference noise in the Peter the Great Bay of the Sea of Japan. The source of the partially coherent broadband interference noise was a seaport and its nearby vessels. Surface wind speed was 5–7 m/s. Wind waves were significant. The receiving system was launched in ~ 47 m of water, with the combined receiver set at a depth of 12 m. Vertical profile of the sound speed is shown in Fig. 4.6, Chap. 4. Axes

x and y of the vector receiver were horizontal, the z axis pointed straight down from the surface to the bottom. The x axis pointed away from the coastline towards the open sea. The launch site was ~ 8 km away from a seaport, with a merchant ship at anchor nearby, and another one ~ 4 km away and cruising slowly in the $+x$ direction towards the system. The vessel carrying the projector was moving in the $-x$ direction. Deployment configuration placed the combined receiving system between two radiation sources. This enabled us to observe compensation of reciprocal flows of tonal energy and interference noise along the x axis. A projector towed 15 m below the surface generated three tone frequencies: 86, 123 and 163 Hz. The realisation time interval was 2400 s long.

3.2.3.2 Experimental Findings

Figures 3.2 and 3.3 show indications of energy from the three tone signals cancelling out the broadband coherent interference noise coming from the seaport. In Fig. 3.2, spectra $S_{p^2}(f)$ and $S_{pV_x}(f)$ are in decibels since $S_{pV_x}(f) = \gamma^2_{pV_x}(f)S_{p^2}(f)$. Signal-to-noise excess in the $S_{p^2}(f)$ spectrum at tone frequencies: 8 dB at 86 Hz; no more than 3 dB at 123 Hz; ~ 12 dB at 163 Hz (Fig. 3.2a). Coherence function $\gamma^2_{pV_x}(f)$ in

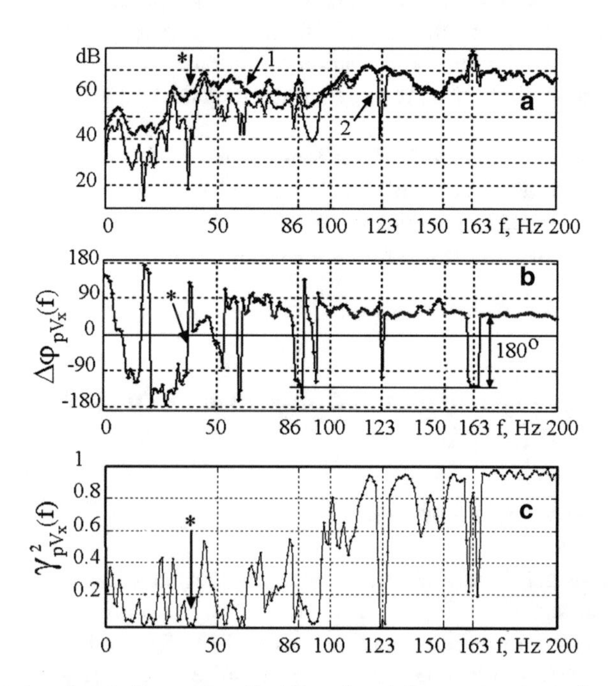

Fig. 3.2 **a** Spectra of acoustic pressure $S_{p^2}(f)$ – 1 and the x component of coherent power $S_{pV_x}(f)$ – 2; **b** x component of the phase spectrum $\Delta\varphi_{pV_x}(f) = \varphi_p(f) - \varphi_x(f)$; **c** spectrum of the x component of frequency coherence function $\gamma^2_{pV_x}(f)$. Averaging time: 3 s. The decibel scale is arbitrary. Tone frequencies: 86, 123, 163 Hz

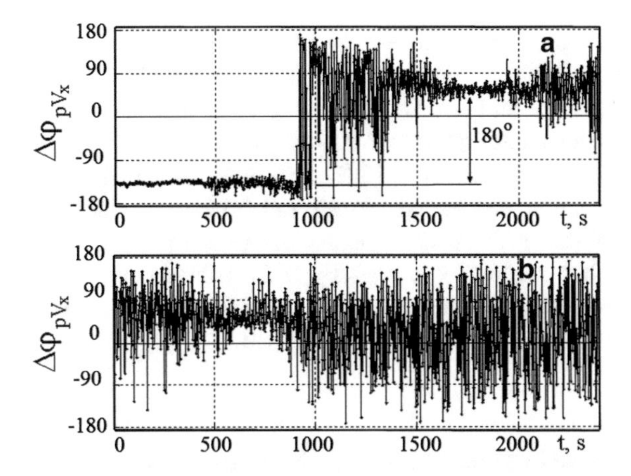

Fig. 3.3 Time dependence of phase difference $\Delta\varphi_{pV_x}(t)$ at two frequencies: **a** $f = 123$ Hz, **b** $f = 36$ Hz. Spectra in Fig. 3.2 correspond to $t = 500$–900 s in this figure. Averaging time: 3 s

Fig. 3.2c is $\gamma^2_{pV_x}(f) \sim 0.2$ at 86 Hz; $\gamma^2_{pV_x}(f) \sim 0$ at 123 Hz; and $\gamma^2_{pV_x}(f) \sim 0.8$ at 163 Hz. Figure 3.2 suggests that the coherence function $\gamma^2_{pV_x}(f)$ and coherent power $S_{pV_x}(f)$ of the interference noise are compensated by the signal field at 123 Hz, and the coherent power $S_{pV_x}(f)$ 'dips' 30 dB at this frequency. Therefore, while spectrum $S_{p^2}(f)$ fluctuates ~ 3 dB at the 132 Hz signal frequency, in the $S_{pV_x}(f)$ spectrum, it can be observed with an 'excess' of (– 30) dB.

The 86 Hz signal is partially compensated in the $S_{pV_x}(f)$ spectrum. The 163 Hz signal is surrounded by coherent noise; signal coherence reaches 0.8, but coherence of interference noise at adjacent frequencies is over 0.9. Phase differences $\Delta\varphi_{pV_x}(f)$ of the broadband interference noise and the 123 Hz tone (Fig. 3.2b) are exactly 180° apart, meaning that signal and noise energy flux density vectors were opposite and the signal completely cancelled out the interference noise, with a 'dip' of – 30 dB.

Coherent power spectrum $S_{pV_x}(f)$ and coherence function $\gamma^2_{pV_x}(f)$ 'dip' at other frequencies as well: for example, at 18 and 36 Hz, associated with the tow vessel. At 36 Hz, the dip reaches 40 dB. This frequency is marked by an asterisk with an arrow in Fig. 3.2. Figure 3.3 shows phase differences $\Delta\varphi_{pV_x}(t)$ at two frequencies (123 and 36 Hz) over a time interval of 2400 s. Spectra in Fig. 3.2 correspond to 500 s $< t <$ 900 s in Fig. 3.3. In this time interval, reciprocal energy flows cancel each other out, but the proper phases of signal and noise are preserved. $\Delta\varphi_{pV_x}(f)$ of signal and noise, being 180° apart, are clearly distinct in Fig. 3.2b. At $t >$ 900 s, $\Delta\varphi_{pV_x}(f)$ jumps 180° (Fig. 3.3a). Fluctuation of averaged phase (averaging time 3 s) at $t >$ 900 s indicates that instantaneous phase oscillates between signal and noise phase.

It is shown in [2–4] that phase difference $\Delta\varphi_{pV_i}(f_0, t)$, where $i = x, y, z$, becomes indefinite at the compensation frequency. As can be seen from Fig. 3.3a, $\Delta\varphi_{pV_x}(t)$ fluctuates relative to the signal phase in the interval $\Delta t_1 = 500$–900 s, meaning that signal phase prevails in the compensation process in this time interval; compensation is also observed in the interval $\Delta t_2 = 900$–2,400 s, but this time the interference phase, which is 180° out of phase with the signal, is the dominant one. Phase differences in the time intervals Δt_1 and Δt_2 are 180°, suggesting that the energy flux vectors of the signal and interference noise arrive from opposite directions.

The degree of compensation at 36 Hz indicates that signal is fully compensated by interference noise; accordingly, $\Delta\varphi_x(t)$ at 36 Hz is indefinite throughout the 2400 s time realisation. Signs of compensation (dips in $S_{pV_x}(f)$ of varying depths) are evident at other frequencies as well (Fig. 3.2). This analysis shows that the excess in coherent processing with compensation can reach 30–40 dB, while signal excess in the acoustic pressure channel is less than 3 dB.

3.3 Compensation of Intensity Over a Broadband of Signal and Dynamic Underwater Acoustic Noise in the Deep Open Ocean

Compensation between broadband signal and noise can be affected by signal interference. The effect of interference on compensation has been observed in deep open ocean environment.

The experiment was conducted aboard R/V *Akademik Lavrentyev* in the Philippine Sea at $\varphi = 18^\circ 53.4'$ N, $\lambda = 126^\circ 38.5'$ E.

3.3.1 Experimental Setup and Technique

Water depth at the observation point was ~ 4900 m. Surface wind speed was ~ 7 m/s. Surface waves were significant. A near-surface homogeneous layer with sound velocities of ~ 1536.8 m/s extended to a depth of 100 m; sound speed reached a minimum of 1482.7 m/s on the axis of the SOFAR channel 1000 m deep; sound speed near the bottom was 1542.0 m/s, which is more than near the surface. In these metocean conditions and under a moderate rain, the spectral level of the z component of ambient noise energy flux $S_{pV_z}(f)$ in the 50–800 Hz range was 60–68 dB re 1 μPa^2/Hz. At a depth of 150 m, the power spectral density of the z component of a local source travelling from the bottom to the surface was comparable with the ambient noise level or lower, depending on whether the frequency was an interference maximum or a minimum. At a depth of 500 *m*, the z component of the spectral power density of the noise-like signal from the local source exceeds that of the vertical component of the signal reflected from the bottom by up to 10 dB in the continuous part of the spectrum. At this depth, refracted energy of the noise-like signal from the vessel and surface noise flows in the same direction.

Figure 3.4 shows the depth profile of the sound speed and the beam pattern of refracted rays SG_1, SG_2, SG_3 and the ray bundle reflected from the bottom bounded

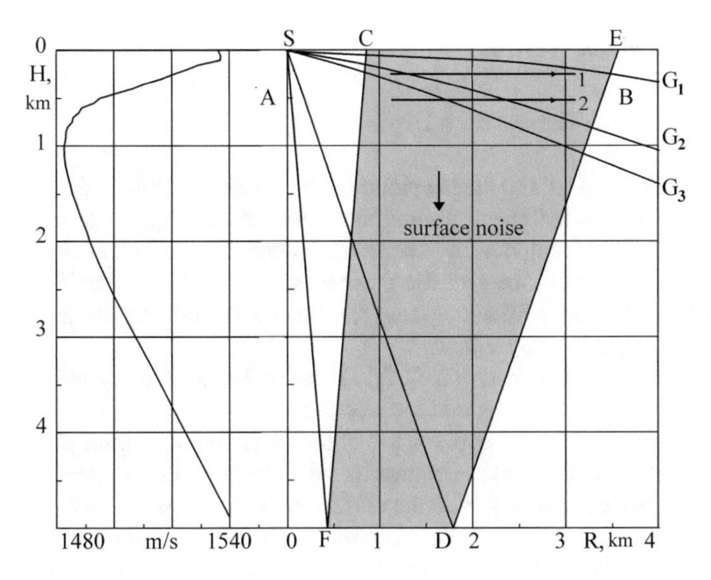

Fig. 3.4 Experimental conditions: **a** sound speed depth profile; **b** layout of combined receivers 1 and 2 relative to the beam pattern of the noise source

by rays SBC and SDE. The horizontal lines represent positions of the two combined receivers of the free-drifting telemetry system during the experiment: 1 at 150 m depth, 2 at 500 m depth.

Coordinate system of the combined receiver: axes x and y are horizontal; z axis points from the surface down to the ocean floor. Figure 3.4 demonstrates that the combined receivers were in the near field of insonification throughout the experiment. Critical refracted ray SG_1, which limits the near field of insonification, exits the source at a 3° angle, ray SG_2 at 10° and ray SG_3 at 15°. Energy flowing from the source S at the observation depth of 150 m will be distributed as follows: horizontal channels x and y of the combined receiver will capture mostly refracted rays with grazing angles of 3–6°; the vertical z channel will capture rays reflected once from the bottom with grazing angles of 70–85°. Refracted rays at 3–6° contribute less than 0.5% of their energy to the vertical energy flow. The second receiver, 500 m deep, is insonified by rays G_2 and G_3 with grazing angles of ~ 20°. These rays contribute ~ 2% to the vertical component. The vertical z channels of the combined receivers also record dynamic surface noise energy propagating downward; this is indicated by a vertical downward arrow in Fig. 3.4b.

The vessel as a noise source manoeuvred as follows. At the beginning of the experiment, the vessel was adrift and silent ~ 1.5 km away from the telemetry system. Then its engine was started, and the noisy vessel passed ~ 1.2 km abeam of the system before pulling away to a distance of ~ 2.5 km and going adrift. The time of recording while the vessel is moving is ~ 7 min. During the experiment, a heavy tropical shower fell over the telemetry system for ~ 10 min, fading into a rain by the time the vessel passed by the receiving system.

3.3.2 Research Results

3.3.2.1 Research Results for a Depth of 150 m

Sonograms in Fig. 3.5 paint the big picture of how scalar and vector acoustic quantities varied over time. In frequency–time coordinates, the sonograms describe acoustic pressure $S_{p^2}(f,t)$, z components of coherence function $\gamma_z^2(f,t)$ and of phase difference $\Delta\varphi_z(f,t)$ between the acoustic pressure and the z component of the acoustic pressure gradient. Phase difference $\Delta\varphi_z(f,t)$ is known to differ from phase difference $\Delta\varphi_{pV_z}(f,t)\Delta\varphi_{pV_z}(f,t)$ by $\pi/2$.

The sonogram of autospectrum $S_{p^2}(f,t)$ reveals distant shipping noise in the 10–70 Hz frequency range throughout the record at ~ 90 dB re 1 μPa²/Hz. The tropical shower is faintly visible at 200 s < t < 500 s in the autospectrum $S_{p^2}(f,t)$. The shower stands in much starker contrast in the z channel of the coherence function compared with the p channel. With the vessel moving (t > 570 s), Fig. 3.5a shows the broadband spectrum of the vessel's noise with discrete lines throughout the spectrum. No interference phenomena are evident in autospectrum $S_{p^2}(f,t)$ or in the sonograms $S_{V_x^2}(f,t)$,$S_{V_y^2}(f,t)$, $S_{V_z^2}(f,t)$,$S_{pV_x}(f,t)$, $S_{pV_y}(f,t)$, $\gamma_{pV_x}^2(f,t)$, $\gamma_{pV_y}^2(f,t)$ [4].

Sonograms $S_{pV_z}(f,t)$, $\gamma_z^2(f,t)$ and $\Delta\varphi_z(f,t)$, on the other hand, display periodic interference changes of coherence and phase difference at t > 750 s. Although sonograms of $S_{pV_z}(f,t)$ are not presented here, $\gamma_z^2(f,t)$ and $\Delta\varphi_z(f,t)$ are the same as $\gamma_{pV_z}^2(f,t)$ and $\Delta\varphi_{pV_z}(f,t)$.

Let us consider individual spectral curves drawn from the body of sonograms (Fig. 3.5b, c). Figure 3.6 shows spectra $\gamma_z^2(f)$ and $\Delta\varphi_z(f)$ of underwater ambient noise recorded at $t = 650$ s. At this time the ship was silent and the shower had eased to moderate rain. The coherence function reaches ~ 0.8 at high frequencies.

$\Delta\varphi_z(f)$ averages to ~ (– 90°), which means that energy of the dynamic surface noise flows from the surface to the bottom [4]. Figure 3.7 shows the spectra of $\gamma_z^2(f)$ and $\Delta\varphi_z(f)$ from sonograms in Fig. 3.5b, c at $t = 890$ s. By that time, the steadily moving vessel was ~ 2 km away from the receiving system.

Let us now consider the relationship among the frequencies that minimise $\gamma_z^2(f)$ (Fig. 3.7). These frequencies are: 14.0; 19.0; 58.0; 63.3; 120.2; 161.9; 205.7; 249.7; 326.5; 403.5; 473.0; 564.5; 600.0; 634.2; 758.7 Hz. They are marked by dots in Fig. 3.7a. At these frequencies, deviations of $\Delta\varphi_z(f)$ are within ± 90° (Fig. 3.7b). Normalising these frequencies to $f_0 = 758.7$ Hz reveals a linear dependence $K_i = f_i/f_0$ (where i runs from 1 to 15) (Fig. 3.7a). If we assume that at these frequencies the relative position of the source and receiver corresponds to an interference maximum of rays reflected from the bottom, then at these frequencies the minima of $\gamma_z^2(f)$ are explained by compensation of two reciprocal energy fluxes: surface noise energy flowing downwards and broadband signal energy reflected from the bottom flowing upwards.

The frequency interval in which the beams interfere is no wider than 3 Hz at each of the frequencies.

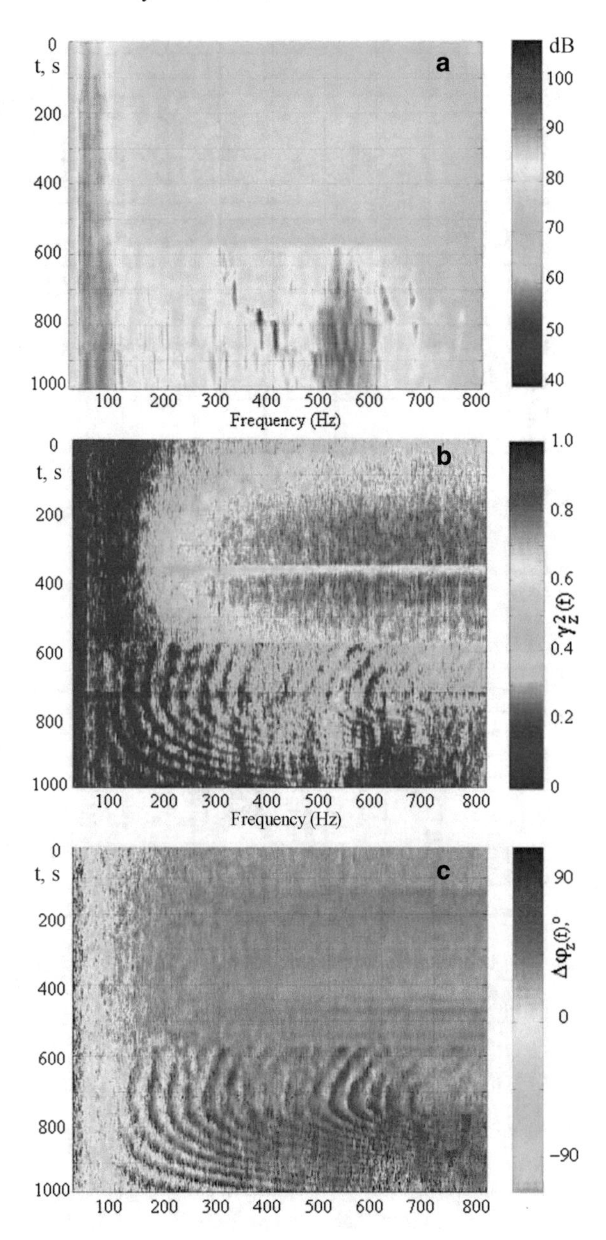

Fig. 3.5 Sonograms of a moving surface source: **a** pressure power $S_{p^2}(f,t)$; **b** z component of coherence function $\gamma_z^2(f,t)$; **c** phase difference between acoustic pressure and the z component of pressure gradient $\Delta\varphi_z(f,t)$. Averaging time: 10 s; analysis band: 1.2 Hz; exponential averaging. Combined receiver depth: 150 m. Vessel start time: 570 s

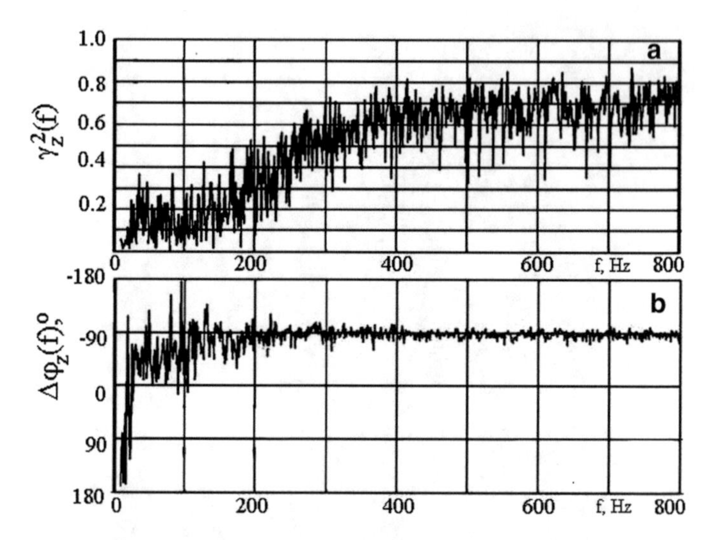

Fig. 3.6 Spectra of the z component of the coherence function **a** and the phase spectrum **b** of underwater ambient noise. The spectra are from Fig. 3.5b, c at $t = 560$ s. Same conditions as in Fig. 3.5

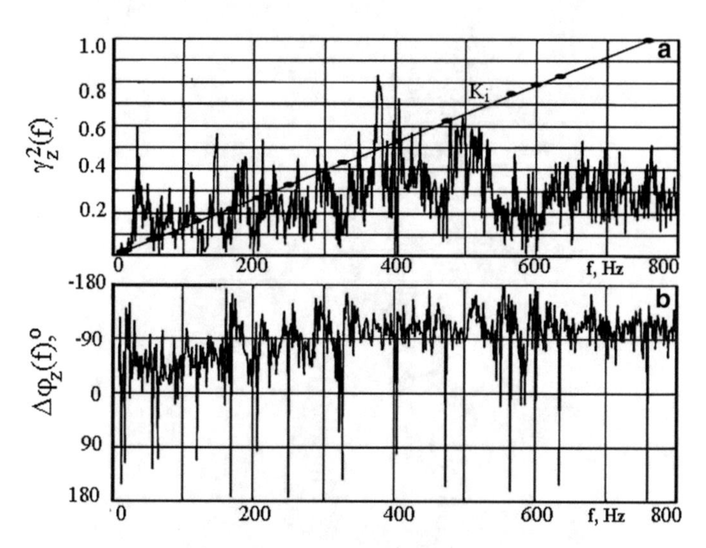

Fig. 3.7 Spectra of the z component of the coherence function **a** and phase spectrum **b** with a moving surface source. Taken from Fig. 3.5b, c at $t = 890$ s. $K_i = f_i/f_0, f_0 = 758.7$ Hz. Same conditions as in Fig. 3.5

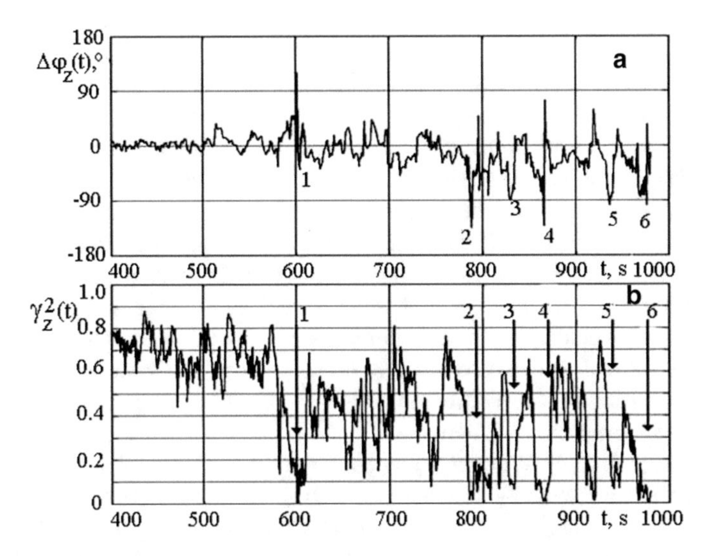

Fig. 3.8 Time dependence in the interval from 400 to 1000 s at $f_0 = 326$ Hz: **a** $\Delta\varphi_z(f_0, t)$, phase difference between acoustic pressure and the z component of particle velocity; **b** $\gamma_z^2(f_0, t)$. Analysis bandwidth: 1.2 Hz. Averaging time: 10 s

Figure 3.8 shows time dependencies of $\Delta\varphi_z(t)$ and $\gamma_z^2(t)$ at 326 Hz in the time interval of 400–1000 s; these are from sonograms Fig. 3.5b, c.

Unlike the previous Figs. 3.6b and 3.7b, in Fig. 3.8 $\Delta\varphi_z(t)$ there is a phase difference between acoustic pressure and the z component of particle velocity.

In Fig. 3.8a, $\Delta\varphi_z(t) = 0$ in the time interval of 400–510 s, which corresponds to the surface noise of the shower; between 510 and 570 s the vessel was preparing to move off with its engine running; at $t > 570$ the vessel was moving in a straight line at a constant speed. The 'leaps' at $t > 570$ s reflect a change in the direction of vertical energy movement. At $\Delta\varphi_z(f) \approx \pm 90°$ (Fig. 3.8a) the reciprocal energy fluxes fully compensate each other (Fig. 3.8b). Figure 3.8 shows six (marked with arrows) of the most obvious 'dips', which occur at $\gamma_z^2(f) \sim 0.1$. Between the 'dips', which is $\gamma_z^2(f)$ ~ 0.7–0.8 and $\Delta\varphi_z(f) \sim 0°$, consistent with the surface noise coherence function in Fig. 3.6.

This experimental result can be described as follows. First consider the formation of signal energy flow reflected from the bottom. Suppose that two coherent plane waves reflected from the bottom arrive at a given observation point. Their path difference is such that maxima or minima of cross-spectral density $S_{pV_z}(f)$ will be observed at certain frequencies as a result of interference. For two such plane waves of the same frequency arriving at the observation point, the average net energy flux density along the z axis will have the form [4]:

$$I_{-z,S} = \frac{1}{2} p_1 V_1 \cos\theta_1 + \frac{1}{2} p_2 V_2 \cos\theta_2 + \frac{1}{2}[p_1 V_2 \cos\theta_2 + p_2 V_1 \cos\theta_1]\cos(\psi_2 - \psi_1), \tag{3.6}$$

where p_1, p_2, V_1, V_2 are pressure and particle velocity amplitudes of the first and second waves, respectively; θ_1, θ_2 are angles that wave vectors of the first and second waves form with the z axis; $(\psi_2 - \psi_1)$ is the phase difference between acoustic pressures or particle velocities of the plane waves; $I_{-z,S}$ indicates that the broadband signal propagates in the z direction from the bottom to the surface.

In (3.6), contributions of the first and the second term to the total energy flux of the signal will be constant. The magnitude and sign of the third term depend on the phase difference $(\psi_2 - \psi_1)$, which varies from 0 to 2π over the wavelength. The upper measurement frequency $f = 800$ Hz is equivalent to wavelength $\lambda \sim 1.9$ m. Because the diameter of the combined receiver is $d \sim 0.2$ m, it follows that $d < \lambda$. It then follows that in our case a single combined receiver is enough to trace the effect of the changing phase difference $(\psi_2 - \psi_1)$ on the z component of intensity at distances far shorter than ~ 1.9 m.

Let us simplify (3.6). Without losing generality, we will consider that $p_1 = p_2$ and $V_2 = V_1$. We can rewrite θ_2 as a sum: $\theta_2 = \theta_1 + \Delta\theta$. Considering $\Delta\theta$ to be small, $\sin \Delta\theta \to 0$ and $\cos \Delta\theta \to 1$. In this approximation, $\cos \theta_2 = \cos \theta_1 \sim 1$. Since the angles $\theta \sim 70$–$85°$ (Fig. 3.4), we assume that $\cos \theta_1 \sim 1$. Expression (3.6) then becomes

$$I_{-z,S} = p_1 V_1 [1 + \cos(\psi_2 - \psi_1)]. \tag{3.7}$$

In a real experiment, the combined receiver measures the difference $I_z(f)$ of two reciprocal energy flows $I_{+z,N}(f)$ and $I_{-z,S}$, $I_{+z,N}(f)$, where $I_{+z,N}(f)$ describes the flow of dynamic surface noise energy in the z direction of the combined receiver:

$$I_z(f) = I_{+z,N}(f) - I_{-z,S}(f). \tag{3.8}$$

We will assume that the average energy flux of surface noise $I_{+z,N}(f_0)$ is constant at a given frequency f_0. For surface noise at frequencies $f_0 > 100$ Hz (Fig. 3.6b), phase spectrum $\Delta\varphi_z(f_0) = 0$ (here, $\Delta\varphi_z(f_0)$ is the phase difference between pressure and particle velocity). Energy flux $I_{-z,S}(f_0)$ resulting from interference of rays reflected from the bottom varies in magnitude with the distance between source and receiver. The measure of change of its magnitude is the phase difference $\delta = (\psi_2 - \psi_1)$. The relationship between δ and phase difference $\Delta\varphi_z$ between pressure and particle velocity in the net total energy flow $I_z(f)$ has the form:

$$\cos \Delta\varphi_z(f) = \frac{1}{2}(1 - \cos \delta). \tag{3.9}$$

It then follows that at $\delta = \pi$ (interference minimum of the noise-like signal) $\cos \Delta\varphi_z(f) = 1$ and $\Delta\varphi_z(f) = 0$, which is immediately apparent in Figs. 3.6b and 3.7b. In Fig. 3.6b this is true for all $f > 100$ Hz, and in Fig. 3.7b for those frequencies at which interference minima occur and the surface noise energy flow prevails.

At $\delta = 0$ (interference maximum) cos $\Delta\varphi_z(f) = 0$ and $\Delta\varphi_z(f) = \pi/2$. At these frequencies in Fig. 3.7, $\gamma_z^2(f)$ reaches its minima and phase difference $\Delta\varphi_z(f)$ jumps accordingly.

Figure 3.8a, b plot $\Delta\varphi_z(t)$ and $\gamma_z^2(t)$ versus time at a frequency $f_0 = 326$ Hz. Here, $\Delta\varphi_z(f_0, t)$ is the phase difference between pressure and the z component of particle velocity. For several values of $\Delta\varphi_z(t)$ numbered 1–5, $\Delta\varphi_z(f_0, t) \to \pm(\pi/2)$- and coherence function $\gamma_z^2(f_0, t) \to 0$. The same pattern is evident in Fig. 3.6. In between the extremes, the net phase difference $\Delta\varphi_z(f)$ of the sum of signal and noise fluctuates near 0°. Note that $\Delta\varphi_z(f) = 0°$ describes the flow of surface noise energy.

To summarise, in the case of compensation of reciprocal energy fluxes, the phase difference between acoustic pressure and particle velocity in the resulting field tends to ($\pm\,\pi/2$), as it does in the case of a deterministic standing wave. This is consistent with the experiment at 326 Hz (Figs. 3.7 and 3.8). The sign is '$\pm$' because the compensation involves a random 'switching' of direction of the net energy flow by 180°. $\Delta\varphi_z(f) = \pi/2$ describes the case when in the net averaged energy flow the change of particle velocity is $\pi/2$ ahead of the change of pressure. When $\Delta\varphi_z(f) = -(\pi/2)$, the change of particle velocity lags behind the change in acoustic pressure by – ($\pi/2$). Like the coherence function, the z component of the cross-spectrum too has 'dips' in its spectrum. The coherence function is a convenient choice because it is the normalised square of the cross-spectrum.

3.3.2.2 Research Results for a Depth of 500 m

Figure 3.9 is a sonogram of $\gamma_z^2(f, t)$ for receiver 2 located at a depth of 500 m. Sonograms for 150 and 500 m depths (Figs. 3.5 and 3.9) are synchronised. An area of high coherence due to the rain is clearly visible on the sonogram at $t < 570$ s. Ambient noise spectra for $\gamma_z^2(f)$ and $\Delta\varphi_z(f)$ at a depth of 500 m are the same as those at 150 m shown in Fig. 3.6.

However, at $t > 570$ s (when the vessel is moving), there are no periodic changes on the sonogram (Fig. 3.9). Spectra $\gamma_z^2(f)$ and $\Delta\varphi_z(f)$ in Fig. 3.10 are rather similar to surface noise spectra (Fig. 3.5).

This is because at a depth of 500 m, grazing angles of rays SG_2 and SG_3 are greater (Fig. 3.4b) than at 150 m. Their contribution to the vertical component of energy flow going from the surface to the bottom is comparable with surface noise at some frequencies and exceeds it at others. For example, at 500–550 Hz the excess is ~ 10 dB. Figure 3.10 illustrates that at some frequencies $\gamma_z^2(f)$ reaches its highest value (~ 1.0). Phase spectrum $\Delta\varphi_z(f)$ (Fig. 3.10b) indicates that in the 50–800 Hz frequency band energy moves in the + z direction—from the surface to the bottom.

Here, energy coming from the vessel and reflected from the bottom cannot compete with the sum of surface noise and noise-like signal flows both travelling from the surface to the bottom. Therefore, if there is no compensation of comparable reciprocal energy flows, then flux spectra $\gamma_z^2(f)$ and $\Delta\varphi_z(f)$ do not have the features indicative of compensation in Figs. 3.7 and 3.8. If reciprocal energy fluxes

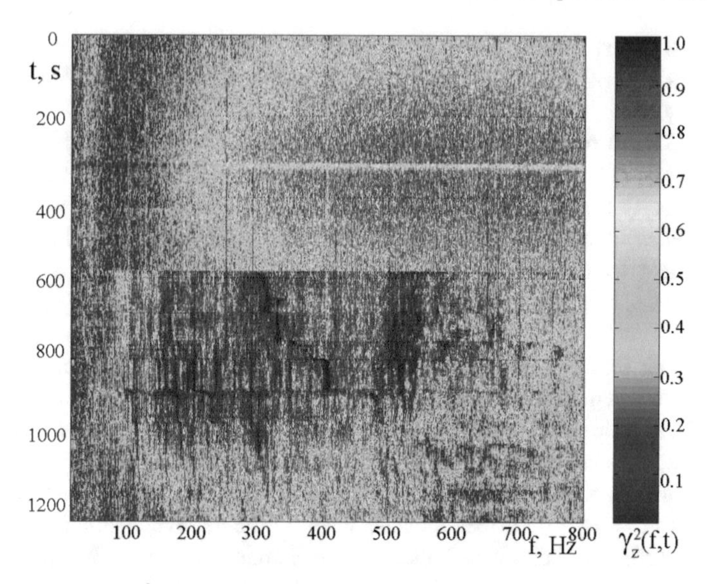

Fig. 3.9 Sonogram of $\gamma_z^2(f, t)$. Measurement depth: 500 m. Same conditions as in Fig. 3.4

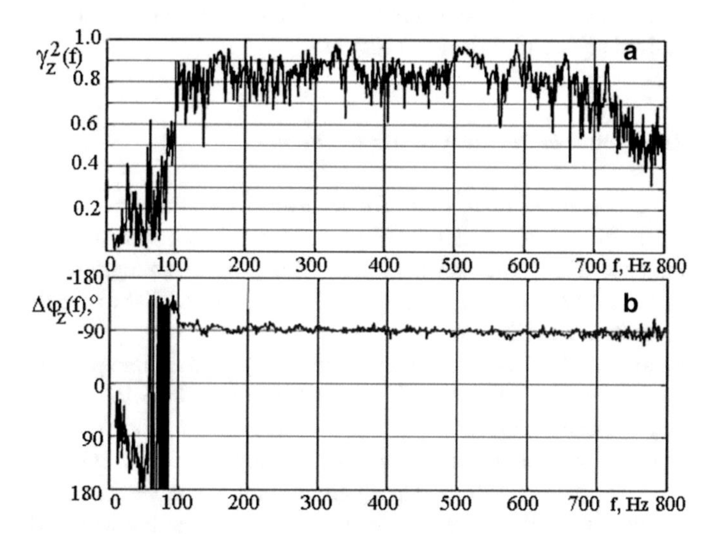

Fig. 3.10 Spectra: **a** $\gamma_z^2(f)$, **b** $\Delta\varphi_z(f)$, the phase difference between acoustic pressure and the z component of pressure gradient. Measurement depth: 500 m, $t = 990$ s. Same conditions as in Fig. 3.5

are observed along the x and y axes, the phenomenon described above is true for these components. In this case, coherence functions (or cross-spectra) should be compared with the corresponding phase spectra.

3.4 Conclusions

Summarising many years of experimental research using vector-phase systems developed by the author [2–6] and based on the analysis in this chapter, we can formulate the following main criteria of occurrence (under certain conditions) of the phenomenon of compensation of reciprocal energy fluxes in underwater acoustic fields.

1. The phenomenon of energy flux compensation can be observed by placing the combined receiver relative to the sound sources in a certain way to detect weak sound sources, thus enhancing the noise immunity of combined receiving systems.
2. Presence of 'dips' in the level of power (coherence) relative to the main spectral level of noise (interference) up to 30 dB deep in cross-spectra $S_{pV_i}(f)$ or in their respective components of the coherence function $\gamma^2_{pV_i}(f)$, where $i = x, y, z$. In acoustic pressure autospectra $S_{p^2}(f)$, signal may exceed noise by up to 3–6 dB at these frequencies (with full compensation).
3. Phase 'jumps' within $\pm(\pi/2)$ observed in the respective phase spectra $\Delta\varphi_x(f)$, $\Delta\varphi_y(f)$ and $\Delta\varphi_z(f)$ at the same frequencies in the case of an interfering signal (Fig. 3.7), or the phase difference becomes indefinite in the case of full compensation (Fig. 3.2).
4. Interfering weak broadband coherent signals travelling in one direction in competition with dynamic surface noise travelling in the opposite direction can indicate a weak sound source. The vertical channel of the combined receiver in the deep ocean can provide the most information for underwater source detection.

References

1. M.A. Isakovich, *General Acoustics* (Nauka, Moscow, 1973). (in Russian)
2. V.I. Ilyichev, V.I. Kuleshov, M.V. Kuyanova, V.A. Shchurov, The interaction of energy flows of underwater ambient noise and a local source. Akusticheskij Zhurnal (Acoustical Physics) **37**(1), 99–103 (1991). (in Russian)
3. V.A. Shchurov, V.I. Ilyichev, V.P. Kuleshov, M.V. Kuyanova, The interaction of energy flows of underwater ambient noise and local source. J. Acoust. Soc. Am. **90**(2), 1002–1004 (1991)
4. V.A. Shchurov, *Vector Acoustics of the Ocean* (Vladivostok, Dalnauka, 2003). (in Russian)
5. L.N. Zakharov, V.I. Ilyichev, S.A. Ilyin, V.A. Shchurov, Vector-phase measurements in ocean acoustics, in *Problems of Ocean Acoustics* (Nauka, Moscow, 1984), pp. 192–204. (in Russian)
6. V.A. Shchurov, V.P. Kuleshov, E.S. Tkachenko, Ivanov E.N, Determinants of compensation of reciprocal energy fluxes in the acoustic fields of the ocean. Akusticheskij Zhurnal (Acoust. Phys.) **56**(6), 835–843 (2010). (in Russian)

Chapter 4
Vortices of Acoustic Intensity Vector in the Shallow Water Waveguide

4.1 Introduction

This chapter presents the author's results in the study of vortices of the acoustic intensity vector in a real shallow water waveguide in 2008–2019. The mechanism behind a local vortex of the acoustic intensity vector (shortened to 'vortex' in what follows) in the far field of a source was first outlined in a series of theoretical papers by Yu. A. Kravtsov and others in 1989–1993. Vortices in the near field of a source have been known for a long time (especially in connection with the problem of equipment noise in air acoustics). Their mechanism in the near field of a source is associated with the phase of particle velocity lagging $\pi/2$ behind acoustic pressure. After Yu. A. Kravtsov's publications (1993) and until 2008, vortices as an object of research disappeared from underwater acoustics research literature. In August 2008, vortices of the acoustic intensity vector in the far field of a source were discovered by the author in an experiment in the Peter the Great Bay of the Sea of Japan. Recent interest in this topical problem can be gauged by the sheer number of publications in Russia and abroad.

The author's research was limited to the frequency range of 20–300 Hz and measurement depths of 15–120 m. Combined vector-phase receiving systems created by the author were used in experimental studies (see Chap. 2).

In the interference vector field surrounding the vortices (outside the separatrix), rotation of the intensity vector must also be nonzero because the lines of the energy flux density vector must envelope the areas where the vortices are. This gives rise to a new property of the vector acoustic field: vortex transfer of energy in the shallow water waveguide. Here, we are talking about large-scale swirling (vorticity) of the vector field. This phenomenon has not been studied enough; full-scale experimental data for the real waveguide are extremely scarce because such studies require advanced vector-phase acoustic technology and appropriate signal processing. The author has thoroughly investigated the phenomenon of vorticity by calculating the rotation of the intensity vector.

V. A. Shchurov, *Movement of Acoustic Energy in the Ocean*,
https://doi.org/10.1007/978-981-19-1300-6_4

This chapter begins with an overview of theoretical work on the mechanism of vortex generation and vortex topology. Presented below are experimental results of studies of tone energy movement in a region of the interference vector field that doesn't contain local vortices, followed by investigation of the spatial interval that contains three vortices. Next, we consider the structure of the vortices and a new phenomenon of mobility of vortices along the horizontal axis of the waveguide in the modal structure; we discover processes of vertical transfer of signal energy that contradict the mode theory. The vortex is considered to be a physical object that is a topological entity with two singular points: a dislocation (centre of the vortex) and a saddle (stagnation point).

Singularity points were first discovered in complex interfering electromagnetic fields in the optical range, which suggests that eddy movement of energy is a universal feature of complex interfering wavefields whatever their nature.

Vortex transfer of mass and energy is ubiquitous in nature. From enormous, awe-inspiring vortex accumulations of mass and energy in the Universe and planetary atmospheres to vortex movements in aero- and hydrodynamics and in acoustic to electromagnetic fields of the optical range—such is the scale of vortex phenomena in nature. Linear dimensions of an acoustic vortex are within half the wavelength of the signal, since the vortex is a product of interference.

Experimental discovery by the Pacific Oceanological Institute team of this fundamental phenomenon in a real acoustic waveguide in 2008 in the Peter the Great Bay blazes new trails in the study of complex acoustic interference fields, as this chapter will demonstrate.

4.2 Fundamental Relationships

Let us consider a list of scalar and vector functions that feature in the energy–momentum tensor and are investigated in this monograph.

4.2.1 Acoustic Pressure, Particle Velocity, Intensity Vector

Let us consider first-order linear acoustic variables. Acoustic field is described by linearised equations of fluid mechanics [1]:

1. Equation of state: $p = c^2\rho + \text{const}$;
2. Euler equation of motion in vector form:

$$\frac{\mathrm{d}\boldsymbol{V}}{\mathrm{d}t} = -\frac{1}{\rho}\mathbf{grad}\,p \tag{4.1}$$

3. Continuity equation: $\operatorname{div}\boldsymbol{V} = -\frac{1}{\rho c^2}\frac{\partial p}{\partial t}$.

To find the wave equation for pressure, we take div of the Euler equation and differentiate the continuity equation with respect to time and subtract one from the other:

$$\text{div}\,\mathbf{grad}\ p = \frac{1}{c^2}\frac{\partial^2 p}{\partial t^2} \quad \text{or} \quad \nabla^2 p = \frac{1}{c^2}\frac{\partial^2 p}{\partial t^2} \tag{4.2}$$

To find the wave equation for the particle velocity in the acoustic wave, we take **grad** of the continuity equation and differentiate the Euler equation with respect to time:

$$\nabla^2 \boldsymbol{V} + \mathbf{rot}\,\mathbf{rot}\,\boldsymbol{V} = \frac{1}{c^2}\frac{\partial^2 \boldsymbol{V}}{\partial t^2} \tag{4.3}$$

Let us write out the Euler equation in terms of components of particle velocity:

$$\frac{\partial V_x}{\partial t} = -\frac{1}{\rho}\frac{\partial p}{\partial x}; \quad \frac{\partial V_y}{\partial t} = -\frac{1}{\rho}\frac{\partial p}{\partial y}; \quad \frac{\partial V_z}{\partial t} = -\frac{1}{\rho}\frac{\partial p}{\partial z}. \tag{4.4}$$

Hence, the components of particle velocity

$$V_x = -\frac{\partial}{\partial x}\int_0^t \left(\frac{p}{\rho}\right)\mathrm{d}t; \quad V_y = -\frac{\partial}{\partial y}\int_0^t \left(\frac{p}{\rho}\right)\mathrm{d}t; \quad V_z = -\frac{\partial}{\partial z}\int_0^t \left(\frac{p}{\rho}\right)\mathrm{d}t. \tag{4.5}$$

Let us introduce the concept of velocity potential, a scalar function of coordinates $\Phi(x, y, z)$:

$$\int_0^t \left(\frac{p}{\rho}\right)\mathrm{d}t = \Phi(x, y, z) \quad \text{or} \quad \boldsymbol{V} = -\mathbf{grad}\ \Phi. \tag{4.6}$$

But $\mathbf{rot}\ \boldsymbol{V} = -\ \mathbf{rot}\ \mathbf{grad}\ \Phi = \mathbf{0}$, meaning that the wavefield of particle velocity is a potential (vortex-free) field. Substituting the velocity potential we find

$$\mathbf{grad}\ p = -\rho\frac{\partial \boldsymbol{V}}{\partial t} = \rho\frac{\partial(\mathbf{grad}\ \Phi)}{\partial t} = \rho\ \mathbf{grad}\frac{\partial \Phi}{\partial t} \quad \text{or} \quad p = \rho\frac{\partial \Phi}{\partial t}, \tag{4.7}$$

which is equivalent to the Euler equation.

Second-order scalars of p and $\boldsymbol{V}$ potential energy density $E_p = \frac{1}{2\rho c^2}p^2$, kinetic energy density $E_k = \frac{1}{2}\rho V^2$ and energy flux density vector (Umov vector) $\boldsymbol{I} = p\boldsymbol{V}$ are related by the acoustic energy conservation law

$$\frac{\partial}{\partial t}\iiint\limits_V\left(\frac{1}{2}\rho V^2+\frac{1}{2\rho c^2}p^2\right)\mathrm{d}v=-\iiint\limits_V \operatorname{div} p\boldsymbol{V}\mathrm{d}v$$
$$=-\iint\limits_S p\boldsymbol{V}\cdot\boldsymbol{n}\mathrm{d}A=-\iint\limits_S \boldsymbol{I}\cdot\boldsymbol{n}\mathrm{d}A \tag{4.8}$$

where p is acoustic pressure, $\boldsymbol{V}$ is the particle velocity vector, c is the speed of sound; ρ is the density of the medium, v is volume and $\boldsymbol{n}$ is the unit normal to the surface A.

The physical meaning of the Umov vector $\boldsymbol{I}$ is instantaneous intensity of the acoustic field. The average of $\boldsymbol{I}$ over the period of a harmonic signal is intensity $\langle\boldsymbol{I}\rangle=\langle p\boldsymbol{V}\rangle$. The unit of the energy flux density (intensity) is $[I]=\frac{\mathrm{J}}{\mathrm{m}^2\mathrm{s}}=\frac{\mathrm{W}}{\mathrm{m}^2}$; the unit of acoustic pressure is $[p]=\frac{\mathrm{H}}{\mathrm{m}^2}=\mathrm{Pa}$ and the unit of particle velocity $[V]=\frac{\mathrm{m}}{\mathrm{s}}$. The acoustic field of the intensity vector may contain a vortex component.

4.2.2 Vector-Phase Characteristics of the Acoustic Field

For purposes of mathematical processing, we consider the signal to be complex and harmonic, and the field stationary and ergodic. To present the acoustic field in complex notation, we introduce the concept of complex intensity $\boldsymbol{I}_c(\boldsymbol{r})$ [2]:

$$\boldsymbol{I}_c(\boldsymbol{r})=\frac{1}{2}p(r)\boldsymbol{V}^*(\boldsymbol{r})=\boldsymbol{I}(\boldsymbol{r})+\mathrm{i}\boldsymbol{Q}(\boldsymbol{r})=\mathrm{Re}I_c(\boldsymbol{r})+i\,\mathrm{Im}I_c(\boldsymbol{r}) \tag{4.9}$$

where $\boldsymbol{I}(\boldsymbol{r})=\mathrm{Re}\boldsymbol{I}_c(\boldsymbol{r})$ is the active and $\boldsymbol{Q}(\boldsymbol{r})=\mathrm{Im}\boldsymbol{I}_c(\boldsymbol{r})$ the reactive intensity vector, $\boldsymbol{r}$ is the spatial variable, and $i=\sqrt{-1}$. If the interference field is made up of many independent waves (rays), then $\mathrm{Re}\boldsymbol{I}_c(\boldsymbol{r})$ and $\mathrm{Im}\boldsymbol{I}_c(\boldsymbol{r})$ are independent random Gaussian functions [3].

For a free field, vector properties of the active $\boldsymbol{I}(r, t)$ and the reactive $\boldsymbol{Q}(r, t)$ intensities can be expressed (in differential form) through rotation and divergence of complex intensity $\boldsymbol{I}_c(r, t)$ for the harmonic signal:

$$\begin{aligned}\mathbf{rot}\,\mathbf{I}_c(r)&=(k/c)\big[(\boldsymbol{I}\times\boldsymbol{Q})/U\big]\\ \mathrm{div}\boldsymbol{I}(r)&=0,\quad \mathbf{rot}\,\boldsymbol{Q}(r)=0,\\ \mathrm{div}\,\boldsymbol{Q}(r)&=2\omega(T-U)=-2\omega L\end{aligned} \tag{4.10}$$

where L is the Lagrangian, $U=\frac{1}{4\rho c^2}p(r)p^*(r)$ is potential energy density, and $T=\frac{\rho}{4}\boldsymbol{V}(r)\boldsymbol{V}^*(r)$ kinetic energy density. From simultaneous Eq. (4.10), it follows that the active intensity vector (energy flux density vector, or the Umov vector) will inherently have vortex properties if $\boldsymbol{I}\times\boldsymbol{Q}\neq\mathbf{0}$—that is, if the vectors $\boldsymbol{I}$ and $\boldsymbol{Q}$ are non-collinear. As full-scale experiment shows, this condition is fulfilled in the shallow water interference field [4–6], even though it is formulated for a free field.

The fundamental significance of this phenomenon, as it follows from (4.10), is that rotation of $\boldsymbol{I}_c(t)$ may be nonzero, more specifically:

$$\begin{aligned}\mathbf{rot}\,\boldsymbol{I}_c &= \mathbf{rot}(p\boldsymbol{V}^*) = p\,\mathbf{rot}\,\boldsymbol{V}^* + [\mathbf{grad}\,p \times \boldsymbol{V}^*]\\ &= [\mathbf{grad}\,p \times \boldsymbol{V}^*],\ \text{since}\ \mathbf{rot}\,\boldsymbol{V}^* = 0.\end{aligned} \tag{4.11}$$

Using the Euler equation $\boldsymbol{V} = -\frac{1}{\mathrm{i}\rho\omega}\mathbf{grad}\,p$ (4.11), we rearrange (4.11) as follows [6]:

$$\begin{aligned}\mathbf{rot}(p\boldsymbol{V}^*) &= -i\frac{\omega\rho}{2}[\mathbf{V} \times \mathbf{V}^*]\\ &= -2\omega\rho\big[V_y V_z \sin(\phi_z - \phi_y)\mathbf{i} + V_x V_z \sin(\phi_z - \phi_x)\mathbf{j}\\ &\quad + V_y V_x \sin(\phi_x - \phi_y)\mathbf{k}\big]\\ &= -2\omega\rho(\mathrm{rot}_x p\boldsymbol{V}^* + \mathrm{rot}_y p\boldsymbol{V}^* + \mathrm{rot}_z p\boldsymbol{V}^*),\end{aligned} \tag{4.12}$$

where $p = P_0 e^{\mathrm{i}(\omega t - \varphi_p)}$ is acoustic pressure; $\boldsymbol{V} = \boldsymbol{V}_0 e^{\mathrm{i}(\omega t - \varphi_v)}$ is the particle velocity vector; $\boldsymbol{V}^*$ is the complex conjugate of the particle velocity vector; ω *is* angular frequency; ρ *is* unperturbed medium density; V_j are amplitudes of particle velocity components ($j = x, y, z$); $(\varphi_z - \varphi_y)$, $(\varphi_z - \varphi_x)$, $(\varphi_x - \varphi_y)$ are phase differences between particle velocity components; and $\boldsymbol{i}, \boldsymbol{j}, \boldsymbol{k}$ are Cartesian unit vectors. From (4.12) it follows that the intensity vector can form a vortex in the far field as well as near the radiation source, provided that at least one of the phase differences is nonzero.

Both scalar, potential energy (U) and kinetic energy (T) play an important role in the field structure: if $T - U = 0$, then div $\boldsymbol{Q} = 0$. In particular, the z component of the reactive intensity vector is related to the acoustic pressure field by the formula $\boldsymbol{Q}_z(r) = -\frac{1}{2\omega\rho}p(r)\mathbf{grad}\,p(r) = -(c^2/\omega)\frac{\mathrm{d}U(r)}{\mathrm{d}z}$ [6].

4.2.3 Energy Streamlines

Energy streamlines geometrically visualise the movement of energy in the acoustic field. In two dimensions (x, y), energy streamlines are integral curves of a first-order differential equation [7]

$$\frac{\mathrm{d}y}{\mathrm{d}x} = \frac{I_y(x, y)}{I_x(x, y)}, \tag{4.13}$$

where $I_y(x, y)$ and $I_y(x, y)$ are the real-valued x and y components of the energy flux density vector

$$\boldsymbol{I}(\boldsymbol{r}) = \frac{1}{2}\mathrm{Re}\,p(\boldsymbol{r})\boldsymbol{V}^*(\boldsymbol{r}) = \frac{1}{2\rho\omega}\mathrm{Im}\,p^*(\boldsymbol{r})\mathbf{grad}\,p(\boldsymbol{r}), \tag{4.14}$$

where $p(\boldsymbol{r})$ is the acoustic pressure, $\boldsymbol{V}(\boldsymbol{r}) = \frac{1}{i\rho\omega}\mathbf{grad}\,p(\boldsymbol{r})$ is the particle velocity vector, ρ is the density of the medium, ω is the angular frequency and $\boldsymbol{V}^*(\boldsymbol{r})$ and $p^*(\boldsymbol{r})$ are complex conjugates of $\boldsymbol{V}(\boldsymbol{r})$ and $p(\boldsymbol{r})$.

One integral curve of the differential equation passes through each point of the (x, y) plane. The streamlines coincide with phase trajectories in the xOy plane—that is, with phase gradient lines. Substituting $p(\boldsymbol{r}) = P(\boldsymbol{r})\exp\{i\Phi(\boldsymbol{r})\}$ in (4.14), we find

$$\boldsymbol{I}(\boldsymbol{r}) = \frac{1}{2\rho\omega}P^2(\boldsymbol{r})\mathbf{grad}\Phi(\boldsymbol{r}) \tag{4.15}$$

From (4.15), it follows that the active intensity vector $\boldsymbol{I}(\boldsymbol{r})$ points in the same direction as phase gradient **grad** $\Phi(\boldsymbol{r})$, meaning that phase trajectories are energy streamlines. We will choose s as a natural parameter on the energy streamline that increases along a unit vector in the direction $\boldsymbol{I}/|\boldsymbol{I}|$. It then follows that on the energy streamline the phase of the wavefield and the natural parameter are related by the differential equality

$$\mathrm{d}\Phi = |\mathbf{grad}\Phi|\mathrm{d}s \tag{4.16}$$

Therefore, the phase of acoustic pressure monotonically increases along the energy streamline. The reactive intensity vector $\boldsymbol{Q}(r) = -\frac{1}{2\rho\omega}P(r)\mathbf{grad}P(r)$ has the same direction as the pressure gradient, meaning that it is normal to the surface of equal pressure. If $\mathbf{I}(r)$ and $\boldsymbol{Q}(r)$ are collinear, then $\mathbf{rot}\,\boldsymbol{I}(r) = \mathbf{0}$ [see (4.10)]. In a plane wave $\exp(i\boldsymbol{kr})$, the streamlines are straight lines parallel to the wave vector. In the vicinity of the singular dislocations and saddles, energy streamlines and phase trajectories take on new properties.

4.2.4 Vortex Generation Mechanism

The interference field of a tone signal in the real shallow water waveguide consists of regions of constructive and destructive interference. Where interference is destructive, acoustic pressure reaches its minima. Energy streamlines on which the absolute value of acoustic pressure tends to zero ('zero lines') and the phase is indefinite form singular points of the phase front: dislocations (centres) and saddles (stagnation points). The concept of 'wavefront dislocation' was introduced by J. F. Nye and M. V. Berry in 1974, by analogy with the notion of dislocation in solid state physics [8, 9]. Figure 4.1 visualises the formation of zero lines with two modes interfering in a waveguide of depth H [10].

The following equations hold at mode intersections in Fig. 4.1a:

$$\begin{cases} |u_1(z)| = |u_2(z)| \\ \cos(\varphi_1 - \varphi_2) = -1 \end{cases} \tag{4.17}$$

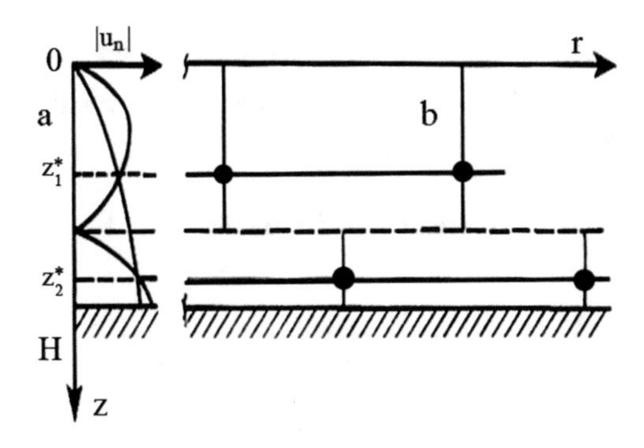

Fig. 4.1 Formation of zeros of acoustic pressure with two propagating modes. The dots (·) are traces of zero lines in the picture plane. The figure is from [10]. Notation in Fig. 4.1: **a** $u_n = a_n \psi_n(z)$, ψ_n are eigenfunctions of the waveguide; a_n is the amplitude of nth mode. We consider the functions $\psi_n(z)$ and a_n to be real-valued; **b** position of pressure zeros (singular zero lines) in the cross-section of the waveguide

Meaning that pressure vanishes if the moduli are equal but the arguments are opposite. Simultaneous Eq. (4.17) suggest that the singular centre points of the vortices are in the (r, z) plane at the intersection of the horizontal lines $z = z^* = \text{const}$, where z^* is the solution of equation (4.17), and vertical segments $\Lambda_{12} = 2\pi / \chi_1 - \chi_2$ apart, where χ_1 and χ_2 are horizontal components of the modes' wave numbers [11]. Zero lines intersect the (r, z) plane at centre points. The singular centre point is an isolated point around which a looped streamline can be drawn. Theoretical works [7, 10, 12] showed that in an ideal waveguide mode interference creates two types of singular points on the wavefront—a dislocation, the centre of a vortex; and a saddle, the point of stagnation—and charted their distribution in the waveguide column. A dislocation has a '+' or a '–' sign; in the centre of a vortex, the vertical component of the energy flux density passes through zero and changes sign; acoustic pressure reaches its minimum values in the centre of a vortex. The point at which separatrix branches intersect is called a saddle. At this point, particle velocity is zero, pressure has a peak, and acoustic pressure and particle velocity are $\pi/2$ out of phase. The separatrix curve limits the area in which the lines of energy movement are looped; energy in a vortex moves between the centre and the saddle towards the source of the signal.

A crucial analytical result was obtained in [12]. For an ideal waveguide with a hard bottom and a soft surface, the paper presents the distribution of centres and saddles in the vertical plane of a 50 Hz harmonic signal waveguide. In a waveguide $H = 150$ m deep, the origin and the horizontal and vertical R and z axes are at the bottom. The potential Φ has the form [11]

$$\Phi = \frac{V_0}{4\pi} j \frac{2\pi}{H} \sqrt{\frac{2}{\pi R}} \exp\left(-j\frac{\pi}{4}\right) \times \sum_{l=1}^{m} \frac{1}{\sqrt{\xi_l}} \cos(b_l z_0) \cos(b_l z) \exp[j(\xi_l R - \omega t)], \tag{4.18}$$

where V_0 is the source strength; ω is the angular frequency of the signal; $\xi_l = \sqrt{k^2 - b_l^2}$ and $b_l = (l - 0.5)\pi/H$ are horizontal and vertical components of the wave vector of the lth mode; m is the number of modes propagating in the waveguide without decay.

Acoustic pressure p and particle velocity components V_R and V_Z can be found from equalities

$$p = -\rho\frac{\partial\Phi}{\partial t}, \quad V_R = \frac{\partial\Phi}{\partial R}, \quad V_z = \frac{\partial\Phi}{\partial z}, \tag{4.19}$$

where ρ is medium density. Components of the energy flux density vector can be found from the equality:

$$\langle I_r\rangle = \frac{1}{2}\langle \mathrm{Re}(pV_R^*)\rangle, \qquad \langle I_z\rangle = \frac{1}{2}\langle \mathrm{Re}(pV_z^*)\rangle. \tag{4.20}$$

Coordinates of the singular centres and saddles can be found from the resulting system of equations, according to [12]. Calculations were made for interference of the 3rd and 4th mode at a distance R of 7,900 to 9,400 m. In Fig. 4.2, centres are marked with ($\cdot$) and saddles with ($\times$).

Where $\cos(\xi_3 - \xi_4)R = \pm 1, \sin(\xi_3 - \xi_4)R = 0$. This paper shows that apart from the criteria $p = 0$ (for a centre) and $|V| = 0$ (for a saddle), another criterion to take into account is when the phase difference between p and V is an odd number of $\pi/2$. In summary, [12] proved with analytical rigour that mode interference creates a topologically stable group consisting of a centre and one or more saddles—a vortex of the acoustic intensity vector.

A more complex vortex pattern results when the interfering modes have substantially different numbers. Paper [12] also considers the case of interference of two

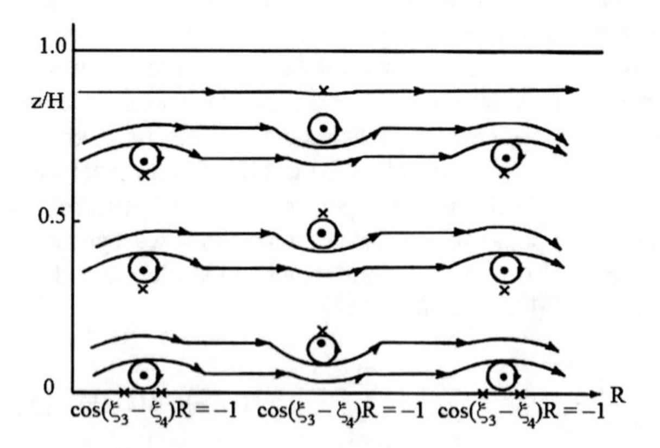

Fig. 4.2 Arrangement of vortices in the case of interference between the 3rd and the 4th modes. Depth $H = 150$ m, frequency 50 Hz. The figure is from [12]

modes with numbers 1 and 10. The structure of the vortices becomes much more complex, which is also noted in [7]. In this case, the energy flux density vector of the signal begins to deviate from the waveguide axis in the region surrounding the vortex on a large scale. This phenomenon is most likely due to strong swirling in the vicinity of the vortices.

4.2.4.1 Models of 'Simple' Vortices

The study of energy transfer in the shallow water waveguide has led to the discovery of a new fundamental physical phenomenon: the vortex of the acoustic intensity vector. The mechanism behind the vortex in the far field of the source is intermodal interference, which points to distinctive features of acoustic energy movement in the region of destructive interference. The study of the vortex structure, which is quite diverse, enables a more subtle analysis of energy processes occurring in the acoustic field. Below, we will profile 'simple' mathematically modelled vortices.

A vortex is a product of interaction of singular points of the wave's phase front—a centre and one or more saddles—and it is topologically stable. Figure 4.3 is a schematic of a vortex formed by two singular points: a centre and a saddle [13].

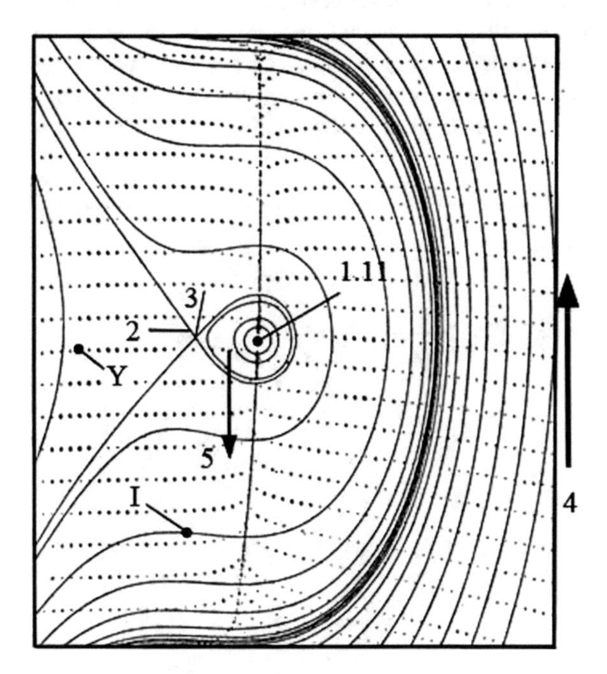

Fig. 4.3 Schematic of a 'simple' vortex with a centre and one saddle [13]. *Legend* 1—centre, 2—separatrix; 3—saddle; $I = \mathrm{Re}\, I_c$; $Y = \mathrm{Im}\, I_c$; 1.11—reactive intensity node. Arrow 4 points where energy is moving from the sound source; arrow 5 shows its movement in the vortex between the centre and the saddle

In Fig. 4.3, energy streamlines are solid lines and reactive component streamlines are dotted lines. While energy streamlines indicate where energy is moving, streamlines of the reactive component indicate how potential energy density of the field is distributed in space. The linear size of the vortex region is ~ 0.2λ. The vortex proper is the area bounded by the separatrix. Energy streamlines that curl around the vortex occupy significant space around it, forming a vorticity space of the energy flux density vector. Whereas on the whole energy moves in the direction of arrow 4 (to the right of Fig. 4.3), in the vortex region between the centre and the saddle energy moves in the opposite direction (arrow 5).

Figure 4.4 depicts the vortex in more detail. The arrows show where energy is moving. Legend: C is the centre of the vortex; B is the saddle. The saddle is at the intersection of the branches of the separatrix. Point C, the centre, is an isolated singularity. Between points B and C, energy flows opposite the overall direction of energy flow from the source.

Figure 4.5 plots normalised values of the following functions along the line passing through points ABCD (Fig. 4.4):

1—active intensity $I(r)$; 2—reactive intensity $Q(r)$, 3—potential energy $E_p(r)$; 4—kinetic energy $E_k(r)$. The figure is from [2]. Figures 4.3, 4.4 and 4.5 show that as energy flows around the vortex (region of point D), the density of energy streamlines increases. Interference redistributes acoustic energy in the area surrounding the local vortex. For example, energy near point D moves faster than it does further away from the vortex. The unit of intensity is $\frac{J}{m^2 s}$. The unit determines the physical meaning of intensity: it is the velocity of transfer of energy density. Energy vortex patterns have been recorded in the near field of several stationary sources [2, 13–16]. In the far

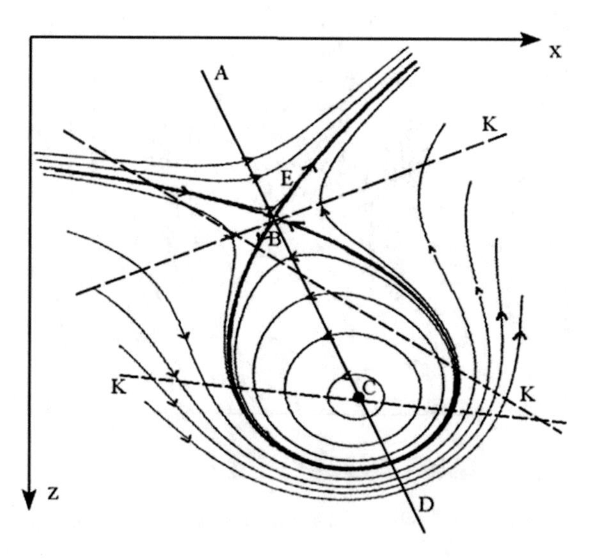

Fig. 4.4 Vortex structure. Arrows on the energy streamlines point in the direction of signal energy transfer

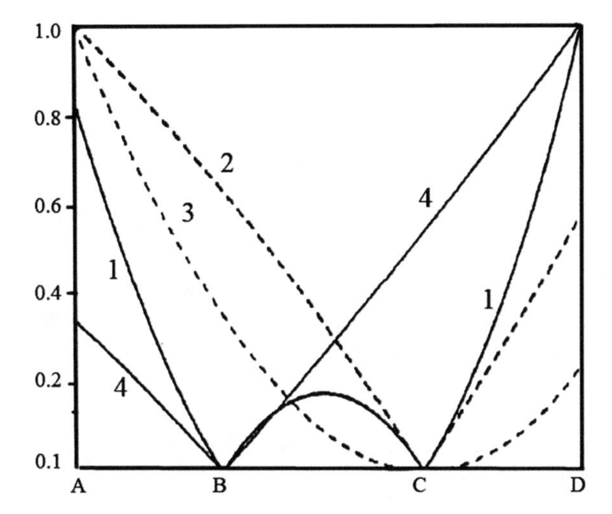

Fig. 4.5 Functions in relation to singular points. *Legend* B—saddle; C—centre. 1—$I(r)$, 2—$Q(r)$, 3—$E_p(r)$, 4—$E_k(r)$ [2]

field of a moving source, positions of vortices have been inferred from anomalous behaviour of scalar and vector functions of the acoustic field [17, 18].

4.3 Vortex Structure of the Interference Field in a Shallow Water Waveguide

4.3.1 Mathematical Processing of Vector Acoustic Signal

To study the dynamics of the interference process that includes constructive as well as destructive areas, we had chosen the following mathematical processing flowchart for the tone signal in the experiment. Mutual statistical processing of experimental time realisations of the four components of the tone field $p(t)$, $V_x(t)$, $V_y(t)$, $V_z(t)$, essentially tantamount to correlation analysis of the data (with a shift $\tau = 0$), relied on FFT in the frequency domain and on the Hilbert transform in the time domain. We investigated autospectra, cross-spectra, differential phase relations, time coherence functions, components of the rotation of active intensity and the grazing angle of energy relative to the waveguide axis.

Phase difference between acoustic pressure and components of particle velocity we find from:

$$\Delta\varphi_{pV_i}(r,\omega) = \arctan\frac{\mathrm{Im}S_{pV_i}(r,\omega)}{\mathrm{Re}S_{pV_i}(r,\omega)},\quad (i = x, y, z) \tag{4.21}$$

and among components of particle velocity $\Delta\varphi_{ij} = \varphi_i - \varphi_j$:

$$\Delta\varphi_{V_iV_j}(r,\omega) = \arctan\frac{\mathrm{Im}S_{V_iV_j}(r,\omega)}{\mathrm{Re}S_{V_iV_j}(r,\omega)}, \quad (i,j = x,y,z), \quad i \neq j \tag{4.22}$$

where r is the spatial variable; $S_{pV_i}(r,\omega)$ is cross-spectral density of acoustic pressure and the ith component of particle velocity; $S_{V_iV_j}(r,\omega)$ is cross-spectral density between the ith and jth components of particle velocity.

The three components of the time coherence function for a given frequency ω_0 from the Hilbert transform we will write as:

$$\gamma_j(t) = \frac{\left\langle \tilde{p}(t)\tilde{V}_j^*(t)\right\rangle_t}{\sqrt{\left\langle \tilde{p}(t)\tilde{p}^*(t)_t\tilde{V}_j(t)\tilde{V}_j^*(t)\right\rangle_t}} = \mathrm{Re}\,\gamma_j(t) + i\,\mathrm{Im}\,\gamma_j(t), \quad j = x,y,z. \tag{4.23}$$

where $\tilde{p}(t)$ and $\tilde{V}_j(t)$ are analytic signals of acoustic pressure and components of particle velocity; i is the imaginary unit in what follows; $j = x, y, z$; $< \ldots >_t$ means linear averaging over several periods of monochromatic signal. Re $\Gamma_j(t)$ and Im $\Gamma_j(t)$ are the normalised values of x, y and z components of the energy flux density: the former is responsible for the transfer of energy in the waveguide, and the latter for the locally bound energy of the field. Variables r and t are equivalent. Expressions (4.21)–(4.23) also hold on average for a random stationary ergodic signal.

4.3.2 *Modes and Vortices*

4.3.2.1 Experimental Setup

To maximise the validity of experimental data, let us consider two experiments in a real waveguide that qualifies as a uniform shallow water waveguide [11, 19]. The experiments were conducted in the Peter the Great Bay of the Sea of Japan.

Measurements were taken with a combined vector-phase receiving systems and a towed projector. The results are from two research cruises in August–September 2013 and 2014. Distances were monitored using the ship's radar.

In the first experiment (August 2013), the combined receiver was 19 m deep in 34 m of water and the projector 14 m deep. The Cartesian axes of the combined receiver were arranged as follows: x and y horizontal, and z vertical from the surface to the bottom. Maximum distance to the vessel was 12,800 m. Figure 4.6a shows that the speed of sound is nearly constant throughout the waveguide water column in the area of study. The radiating vessel was moving towards the receiving system at a constant speed of 2 knots (1 m/s). During the tow, the projector was always in

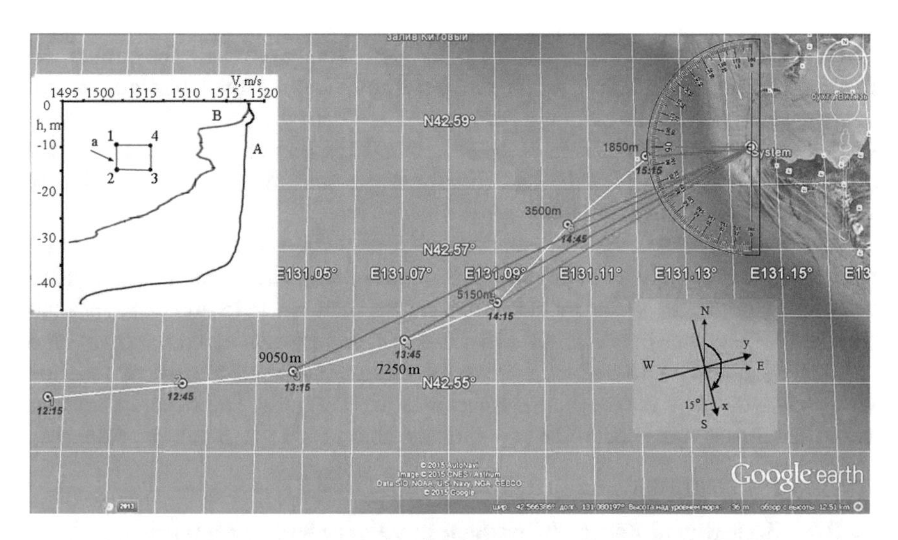

Fig. 4.6 163 Hz audio source towing chart. The inset above shows sound speed profiles: A—first experiment, 163 Hz radiation; B—second experiment, 88 Hz radiation. a—position of the combined system in the waveguide. Rectangle 1–4 is not drawn to scale. The inset below shows how axes of the vector receiver $x0y$ were oriented relative to N. The receiver is at the point marked 'system'

the fourth Cartesian quadrant x-y of the combined receiver. The signal energy flux density vector formed an $\psi \leq 10°$ angle with the y axis. A tone was emitted at a frequency $f_0 = 163$ Hz.

The towing lasted from 12:15 to 15:15 h local time. The distance travelled is easy to work out because the towing speed averaged 1 m/s. Two towing intervals were chosen for experimental data analysis. Each time realisation lasted 1700 s; the distance travelled was 1700 m. In the first time interval (13:15–13:45 h), the distance from source to receiver varied from ~ 9050 to ~ 7250 m; in the second (14:45–15:15 h), from ~ 3650 to ~ 1850 m. Figure 4.6 shows thc towing chart of the 163 Hz sound source.

In the second experiment (September 2014), the receiving module consisted of four combined receivers all enclosed in a common housing displacing ~ 3 m^3.

Each receiver also had its own separate housing. A dual-link suspension and a significant added mass of the receiving module shielded the receiver from vibrations and flow noise (interference noise). Each combined receiver is a three-component vector receiver 180 mm in diameter with six hydrophones mounted on its body. The receivers formed a rectangle in the vertical plane. Receivers 1–4 and 2–3 were 1.2 m apart horizontally; receivers 1–2 and 3–4 were spaced 0.67 m vertically (Fig. 4.6a). No diffraction effect of receivers on each other was detected at a wavelength $\lambda \sim$ 17 m.

The Cartesian axes of the x and y channels of all vector receivers were horizontal and oriented in the same way; the z channel axes were vertical and pointed from the surface to the bottom of the waveguide. A frequency of 88 Hz was emitted. The ship was moving in a straight line at 1.5 m/s. The towing time interval was 200 s. The distance between source and receiver varied from ~700 to ~400 m. The emitter was in the third Cartesian quadrant of the receiver throughout the tow. The signal energy density vector formed an azimuthal angle $\psi \leq 15°$ with the x axis. The experiment was carried out in the same area of the Peter the Great Bay of the Sea of Japan. The bottom of the waveguide is flat and slightly slopes towards the mouth of the bay, with depths in the 30–42 m range.

Bottom sediments in the area consist of variously sized sands: coarse sand in the surface stratum and gravel to pebbles in the second and third strata. Compressional wave velocities in the sediment stratum averaged between 1575 and 1810 m/s, and shear wave velocities between 300 and 475 m/s. The sediment stratum is 50 m thick or less [20]. Based on the bottom's geological structure, we consider the real shallow water waveguide to meet the criteria for a uniform waveguide. The weather remained consistent throughout each of the experiments, with winds ~ 2–3 m/s and calm seas.

4.3.2.2 Tone Signal Energy Movement in a Real Shallow Water Waveguide

First experiment, $f_0 = 163$ Hz. No vortices were detected at the receiver depth in this experiment.

The movement of tone energy in the waveguide was studied by analysing energy and phase characteristics of the acoustic field versus time and distance. Tone frequency $f_0 = 163$ Hz, wavelength $\lambda = 9.3$ m at sound speed $c_0 = 1520$ m/s. In this experiment, the azimuthal angle $\psi(t)$ of the energy flux density vector with the y axis in the horizontal plane is 0°–10°. The y axis is the horizontal axis of the actual waveguide. We will only consider the y components of the functions. Figure 4.6 shows the tow diagram.

We investigated the following functions of time and distance that figure in the energy–momentum tensor: envelope of spectral power density of acoustic pressure $S_{p^2}(t)$ and envelopes of power of orthogonal components of the particle velocity vector $S_{V_x^2}(t)$, $S_{V_y^2}(t)$, $S_{V_z^2}(t)$; phase difference $\Delta\varphi_{pV_j}(t) = \varphi_p(t) - \varphi_{V_j}(t)$; real and imaginary parts of time coherence $\gamma_j(t) = \text{Re}\,\gamma_j(t) + i\,\text{Im}\,\gamma_j(t)$; azimuthal angle $\psi(t)$ and grazing angle of the energy flux $\theta(t)$, where $j = x, y, z$.

The time t and distance r variables are considered equivalent. For convenience, we will use 't' for the towing time. The time t is counted in seconds from the start of the experiment.

Consider the movement of tone signal energy along the horizontal axis of the waveguide at the reception point over time. Figure 4.7 presents the results of statistical processing of the functions $S_{p^2}(t)$, $S_{V_y^2}(t)$, $\Delta\varphi_{pV_y}(t)$ and $\text{Re}\,\gamma_y(t)$ in the first time interval 13:15–13:45 h. At $t_1 = 2100$ s, the distance from the source to the receiver is ~ 9000 m; at $t_2 = 3800$ s, the distance is ~ 7200 m. Data averaging time $\Delta t = 1$ s. Spatial averaging interval is 1 m. The peaks of $S_{p^2}(t)$ and $S_{V_y^2}(t)$ increase with time over the 1700 m interval. The envelope curves of pressure $S_{p^2}(t)$ and the y components of particle velocity $S_{V_y^2}(t)$ are similar; phase difference $\Delta\varphi_{pV_y}(t)$ fluctuates about zero; $\text{Re}\,\Gamma_y(t) = +1.0$ (Fig. 4.7). From this, it follows that $p(t)$ and $V_y(i)$ are in phase, and

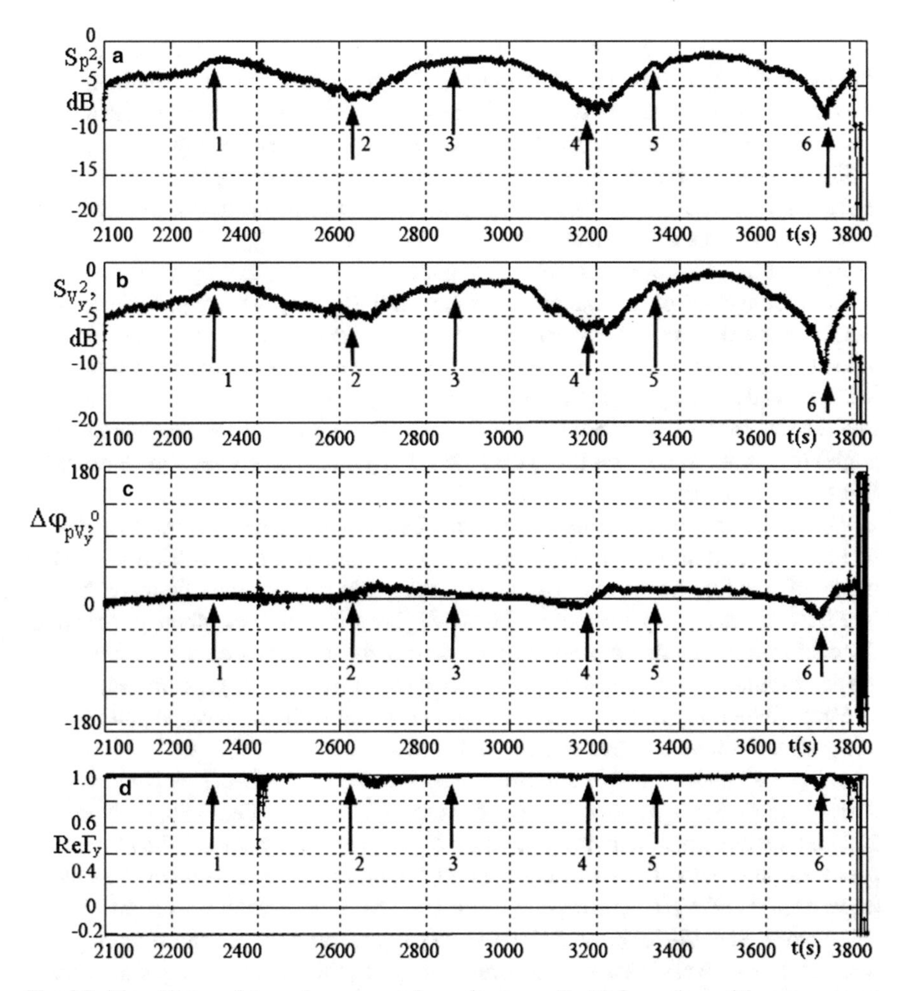

Fig. 4.7 Time (distance) dependence: **a** envelope of pressure $S_{p^2}(t)$; **b** envelope of the y component of velocity $S_{V_y^2}(t)$; **c** phase difference $\Delta\varphi_{pV_y}(t)$; **d** y component of the real part of coherence $\mathrm{Re}\,\Gamma_y(t)$. $f_0 = 163$ Hz. Averaging time $\Delta t = 1$ s

a coherent plane wave travels in the $+\,y$ direction (along the waveguide's horizontal axis) at the point of observation. If so, the relationship between pressure $p(t)$ and the y component of particle velocity $V_y(t)$ obeys the formula $p(t) = \rho\gamma V_y(t)$, where γ is the phase velocity of the normal mode [1]. Experimentally, this has also been noted in [6, 21].

Figure 4.7 indicates that there are no intensity vector vortices at this depth in the waveguide. If there were any, the following functions in Fig. 4.7 would have anomalies near the vortex: phase difference $\Delta\varphi_{pV_y}(t)$ would skip a multiple of 2π; $\mathrm{Re}\,\Gamma_y(t)$ would equal – 1.0 when the phase jumps. That these requirements are not met means there are no vortices of the intensity vector at this depth in the waveguide

[7, 17]. Minor interference noise on the $\Delta\varphi_{pV_y}(t)$ and Re $\Gamma_y(t)$ curves is most likely due to technical reasons. The experimental $S_{p^2}(t)$ curve is similar to the theoretical pressure curve in the case of two interfering close modes. In this case, we are dealing with an acoustic interference field with a spatial period $\Lambda_{12} = 2\pi/\Delta\varkappa_{12}$ equal to ~ 550 m (time interval ~ 550 s). The difference between horizontal components of wave numbers $\Delta\varkappa_{12} = 10^{-2}$ m^{-1}. Altogether, the experiment fully agrees with the mode theory in the case of a uniform (ideal) waveguide [11].

Now let us consider the movement of energy along the *z axis* by analysing these functions: envelope of spectral power density of the *z* component of particle velocity $S_{V_z^2}(t)$; phase difference between acoustic pressure and the *z* component of particle velocity $\Delta\varphi_{pV_z}(t) = \varphi_p(t) - \varphi_z(t)$; real part of the *z* component of time coherence Re $\Gamma_z(t)$ and its imaginary part Im $\Gamma_z(t)$; grazing angle $\theta(t)$ of the energy streamline relative to the *y* axis in the vertical plane *y*0*z* (the *y* axis corresponds to a 90° angle) (Fig. 4.8). $S_{V_z^2}(t)$ is ~ (5–6) dB below $S_{V_y^2}(t)$; the two curves appear different (Figs. 4.7b and 4.8a).

$S_{V_z^2}(t)$ rises slightly and linearly in time; its fluctuations never exceed 1.5 dB.

Phase difference $\Delta\varphi_{pV_z}(t)$ must equal $\pi/2$ throughout the time realisation because a field of standing waves must exist along the *z* axis according to the mode theory. In Fig. 4.8b, $\Delta\varphi_{pV_z}(t)$ fluctuates about $\pi/2$, but these are not inherently random fluctuations that can be suppressed by increasing the averaging time Δt; rather, they are deterministically related to the envelope of spectral density of pressure (Fig. 4.7a).

Points where $\Delta\varphi_{pV_z}(t) = \pi/2$ strongly correlate with the maxima and minima of $S_{p^2}(t)$ and singular points of other functions.

In Figs. 4.7 and 4.8, this relationship is represented by numbers 1–6. Numbers 1–6 correspond to $\Delta\varphi_{pV_z}(t) = \pi/2$ and therefore Re $\Gamma_z(t) = 0$ (Fig. 4.8b, c). Numbers 1, 3 and 5 enumerate the points where Re $\Gamma_z(t)$ changes sign from '–' to ' + ' as it passes through zero. They are near the highs of *p*(*t*) and $V_y(t)$. Numbers 2, 4 and 6 (lows of *p*(*t*) and $V_y(t)$) correspond to Re $\Gamma_z(t)$ passing through zero from ' + ' to '–'. According to [5–7, 18, 22, 23], Re $\Gamma_z(t)$ passing through zero and changing sign with *p*(*t*) at its lowest is a sign of a vortex. However, the absence of signs of vortices (Fig. 4.7) for $\Delta\varphi_{pV_y}(t)$ and Re $\gamma_y(t)$ indicates that there are no vortices at this depth.

The physical meaning of Re $\Gamma_z(t)$ is normalised energy flux in the *z* direction. When $\Delta\varphi_{pV_z} < \pi/2$, Re $\Gamma_z(t) > 0$ and energy flows from the surface to the bottom in the + *z* direction; when $\Delta\varphi_{pV_z} > \pi/2$, Re $\Gamma_z(t) < 0$ and $\Delta\varphi_{pV_z}(t)$ energy flows up in the – *z* direction (4.21).

Let us rewrite $\Delta\varphi_{pV_z}(t)$ as a sum of two terms: $\pi/2$ and a fluctuating residual $\pm\ \alpha(t)$: $\Delta\varphi_{pV_z}(t) = \pi/2\pm\alpha(t)$. In a standing wave along the *z* axis, the following must be true: Re $\Gamma_z(t) = 0$, Im $\Gamma_z(t) = 1.0$, $\Delta\varphi_{pV_z}(t) = \pi/2$ and $\alpha(t) = 0°$. These requirements of the theory are only partially satisfied. Fluctuations $\alpha(t)$ occasionally reach $\alpha(t) \leq\ \pm 45°$ (Fig. 4.8b). Meanwhile, Re $\Gamma_z(t)$ ranges from $-$ 0.4 to + 0.6. This indicates an alternating flux of signal energy in the vertical plane along the vertical *z* axis on top of an existing system of standing waves. In areas (numbers 1–6) where Re $\Gamma_z(t)$ is zero or near zero there is no vertical energy flow. In these areas, there is no transfer of energy along the *z* axis and $\Delta\varphi_{pV_z}(t) = \pi/2$, as in a standing wave. This is fully consistent with the behaviour of Im $\Gamma_z(t)$ (Fig. 4.8d).

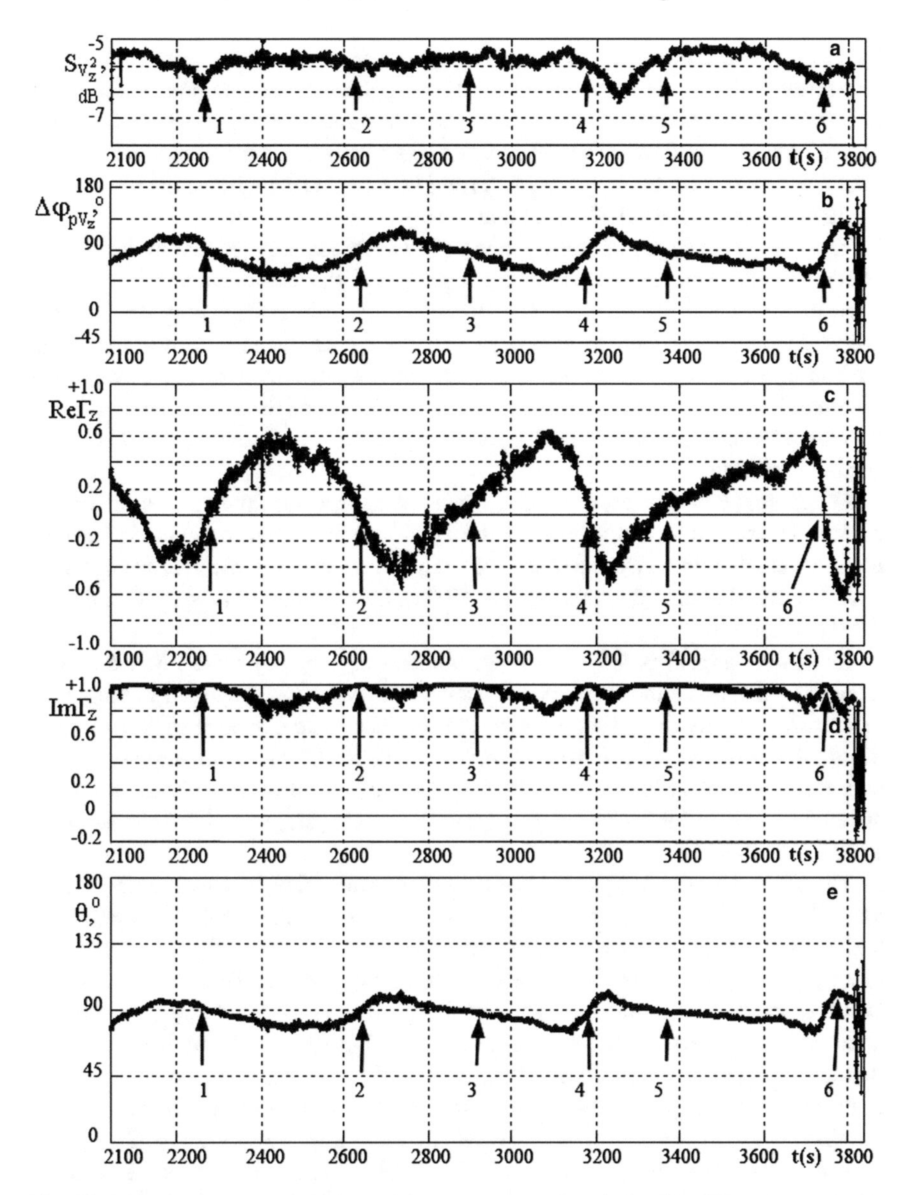

Fig. 4.8 Time dependence: **a** envelope of the z component $S_{V_z^2}(t)$; **b** phase difference $\Delta\varphi_{pV_z}(t)$; **c** real part of z component Re $\Gamma_z(t)$; **d** z component of the imaginary part of coherence Im $\Gamma_z(t)$; **e** grazing angle $\theta(t)$. $f_0 = 163$ Hz. Averaging time $\Delta t = 1$ s

Im $\Gamma_z(t) = +\,1.0$ indicates no signal energy transfer along the z axis. On the time realisation 1700 s long (over a distance of 1700 m), Im $\Gamma_z(t) = +\,1.0$ much of the time. However, the requirements of the theory are not satisfied everywhere, since quasi-periodic alternating flows of signal energy uncompensated on the z axis develop. The fluctuation period of $\Delta\varphi_{pV_z}(t)$, Re $\Gamma_z(t)$ and $\theta(t)$ is ~ 560 s—almost the same as the ~ 550 s fluctuation period of $S_{p^2}(t)$ and $S_{V_y^2}(t)$.

The movement of signal energy along the waveguide axis (y axis) in the vertical plane $y0z$ is determined by the grazing angle $\theta(t)$ of the energy streamline. Grazing angle $\theta(t)$ is measured from the z axis. $\theta(t) = 90°$ means that the energy streamline is horizontal. Net component of intensity $\boldsymbol{I}_{yz} = \mathbf{j}I_y(t) + \mathbf{k}I_z(t)$ in the vertical plane $y0z$, tangent to the energy streamline, quasiperiodically deviates from the waveguide's horizontal axis. In this case, these deviations don't exceed $\pm 15°$. To summarise,

This experiment shows that the requirements of the normal mode theory are partially fulfilled.

Let us now consider experimental results of the second time realisation 14:45–15:15h. The radiation source is still in the fourth quadrant of the combined receiver's coordinates, this time closer to the y axis. The distance to the receiver has changed significantly: this time it varies from ~ 3500 m (at $t = 8000$ s) to ~ 1850 m (at $t =$ 9800 s) (Fig. 4.6).

The functions in Figs. 4.9 and 4.10 are the same as in Figs. 4.7 and 4.8. It is evident from Fig. 4.9a, b that the envelopes of acoustic pressure $S_{p^2}(t)$ and the y component of particle velocity $S_{V_y^2}(t)$ are similar and in phase. The envelope of $S_{V_z^2}(t)$ has a different dependence on t. Levels of curves $S_{p^2}(t)$, $S_{V_y^2}(t)$,$S_{V_z^2}(t)$ increase by ~ 5 dB over a distance of 1700 m. As in the first experiment, no vortex-related anomalies are evident in Fig. 4.9 because $\Delta\varphi_{pV_y}(t) \approx 0°$ and Re $\gamma_y(t) = +\,1.0$ for most of the 1700 s time interval.

Therefore, signal energy is carried by a plane wave along the waveguide axis (y axis). Anomalies of $\Delta\varphi_{pV_y}(t)$ and Re $\gamma_y(t)$ are observed in intervals 8000–8100 and 9200–9400 s, most likely caused by vortices at other depths [18]. The movement of energy along the z *axis* is identical to the first experiment (Figs. 4.8 and 4.10). Points of interest are numbered 1–5. As follows from Fig. 4.10, $\Delta\varphi_{pV_z}(t)$ fluctuates about $\pi/2$. The fluctuations reach $\pm\,\pi/4$. Comparing Figs. 4.8b and 4.10a suggests that $\Delta\varphi_{pV_z}(t)$ experiences longer fluctuations. But the nature of the phenomenon is the same in both cases. As in the first instance, we will write out $\Delta\varphi_{pV_z}(t)$ in the form $\Delta\varphi_{pV_z}(t) = \pi/2 \pm \alpha(t)$. Numbers 1–5 correspond to $\Delta\varphi_{pV_z}(t) = \pi/2$, Im $\gamma_z(t) = +1.0$,Re $\gamma_z(t) = 0$. As before, numbers 1, 3 and 5 mark the highs of $S_{p^2}(t)$—at these points, Re $\Gamma_z(t)$ passes through zero and changes sign from '–' to ' + '; numbers 2 and 4 correspond to the lows of $S_{p^2}(t)$—there, Re $\Gamma_z(t)$ passes through zero and changes sign from ' + ' to '–'. The fluctuation period of $\Delta\varphi_{pV_z}(t)$, Re $\Gamma_z(t)$ and $\theta(t)$ is 665 s; for $S_{p^2}(t)$ and $S_{V_y^2}(t)$, the fluctuation period is ~ 588 s. Interference curves $S_{p^2}(t)$ and $S_{V_y^2}(t)$ in Fig. 4.9a are the product of interference of modes beyond the first few, since the distance between the projector and the receiver has significantly shortened. Nevertheless, fluctuation periods in the first and second cases are comparable.

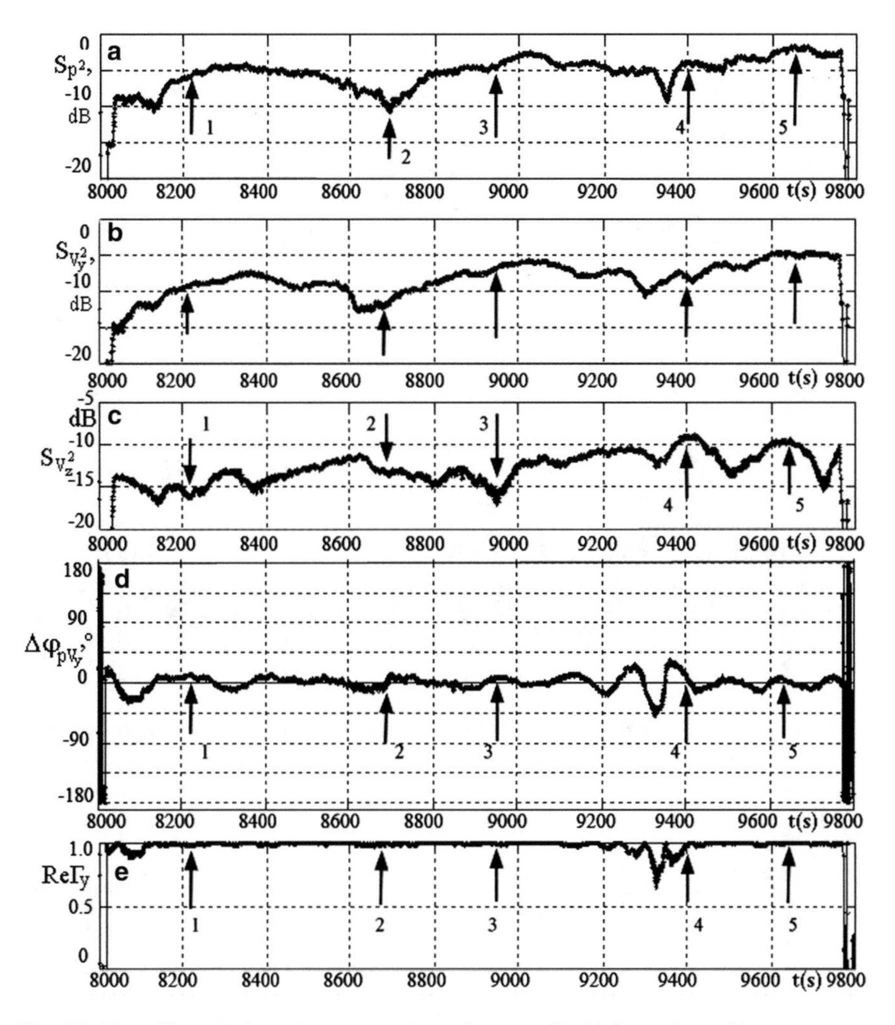

Fig. 4.9 Time (distance) dependence: **a** envelope of pressure $S_{p^2}(t)$; **b** envelope of the y component of velocity $S_{V_y^2}(t)$; **c** envelope of the z component of velocity $S_{V_z^2}(t)$; **d** phase difference $\Delta\varphi_{pV_y}(t)$; **e** real part of the y component $\mathrm{Re}\,\Gamma_y(t)$. $f_0 = 163$ Hz. $f_0 = 163$ Hz. Averaging time $\Delta t = 1$ s

The alternating function $\mathrm{Re}\,\Gamma_z(t)$ ranges within $\pm$ 0.8 with a spatial period of ~ 665 m. Between numbers 1 and 5 the imaginary part $\mathrm{Im}\,\Gamma_z(t)$ reaches + 1.0 (at the points where $\mathrm{Re}\,\Gamma_z(t) = 0$) and up to ~ 0.8 (where $\mathrm{Re}\,\Gamma_z(t) \neq 0$). This means that a standing wavefield is observed at and near points 1–5 from time to time. Uncompensated flows of signal energy along the z axis distort this pattern to a greater or lesser degree. The grazing angle $\theta(t)$ of the signal energy flow fluctuates about the horizontal axis of the waveguide. The farthest $\theta(t)$ deviates from the horizontal towards the bottom is 45° (Fig. 4.10d). Fluctuations of $\boldsymbol{I}_{yz} = \mathbf{j}I_y(t) + \mathbf{k}I_z(t)$ about the horizon of the waveguide are stronger than in the first case.

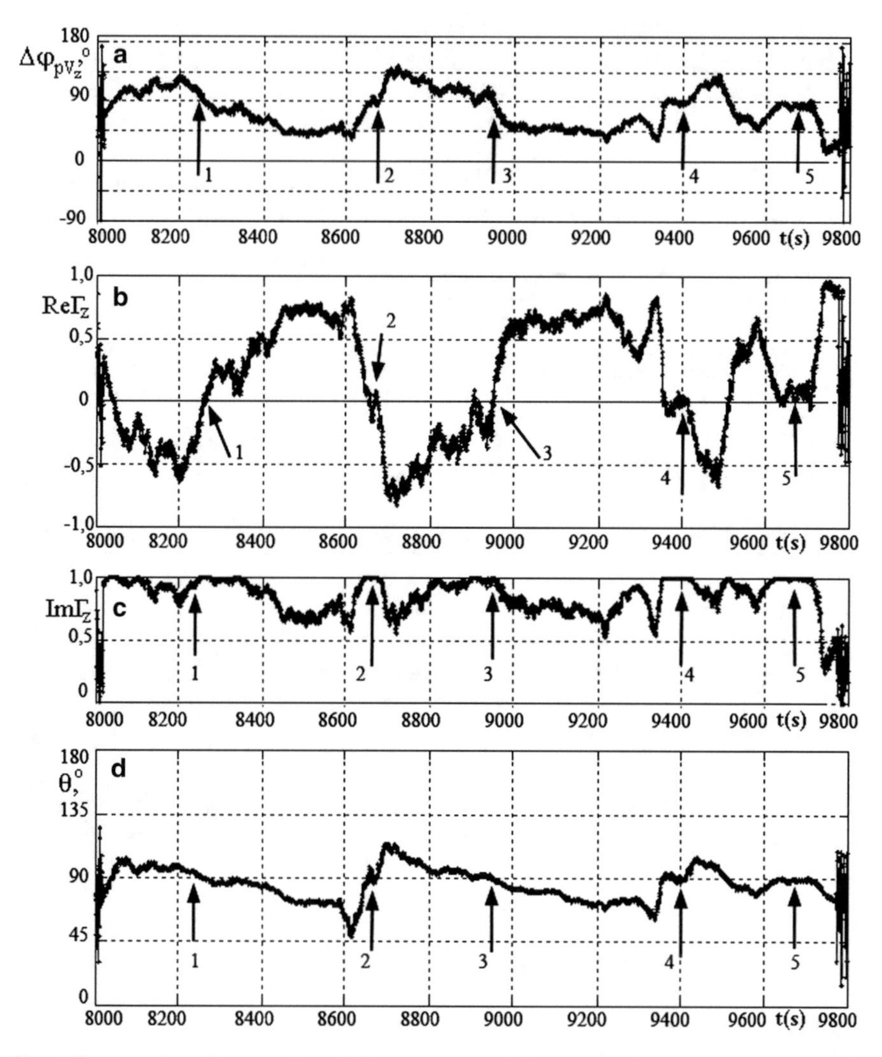

Fig. 4.10 Time dependence: **a** phase difference $\Delta\varphi_{pV_z}(t)$; **b** real part of the z component Re $\Gamma_z(t)$; **c** imaginary part of the z component Im $\Gamma_z(t)$; **d** grazing angle $\theta(t)$. $f_0 = 163$ Hz. Averaging time $\Delta t = 1$ s

To summarise, in the far field of the source in the absence of local vortices at the receiver depth in the shallow water waveguide, the experiment fits the mode theory along the horizontal axis of the waveguide. In the vertical plane along the z axis, alternating vertical fluxes of signal energy periodically occur alongside a standing wavefield. The mechanism of this phenomenon has to do with how far the source is from the receiver. The effect of underwater ambient noise on the experimental result is negligible because signal-to-noise ratio is 15–20 dB.

Second experiment, $f_0 = 88$ Hz. The radiating vessel moves in a straight line towards the receiving system in the range of ~ 700 to ~ 400 m at a speed of 1.5 m/s. The radiation source is in the receiver's third Cartesian quadrant. The signal intensity vector forms an azimuthal angle of ~ 15° with the x axis; emission frequency is 88 Hz, $\lambda = 17$ m and $c = 1510$ m/s. The functions investigated are the same as in the first experiment. Averaging time is 1 s. Spatial averaging interval is 1.5 m. The x axis points along the horizontal axis of the waveguide.

Figure 4.11 shows the envelopes of three components of spectral power density of the acoustic field: $S_{p^2}(t)$, $S_{V_x^2}(t)$ and $S_{V_z^2}(t)$. All y characteristics are identical to their x counterparts and so are not shown here. As the vessel approaches the receiver, the spectral level of acoustic pressure and particle velocity increases by ~ (5–6) dB around the highs and by ~ 2 dB around the lows.

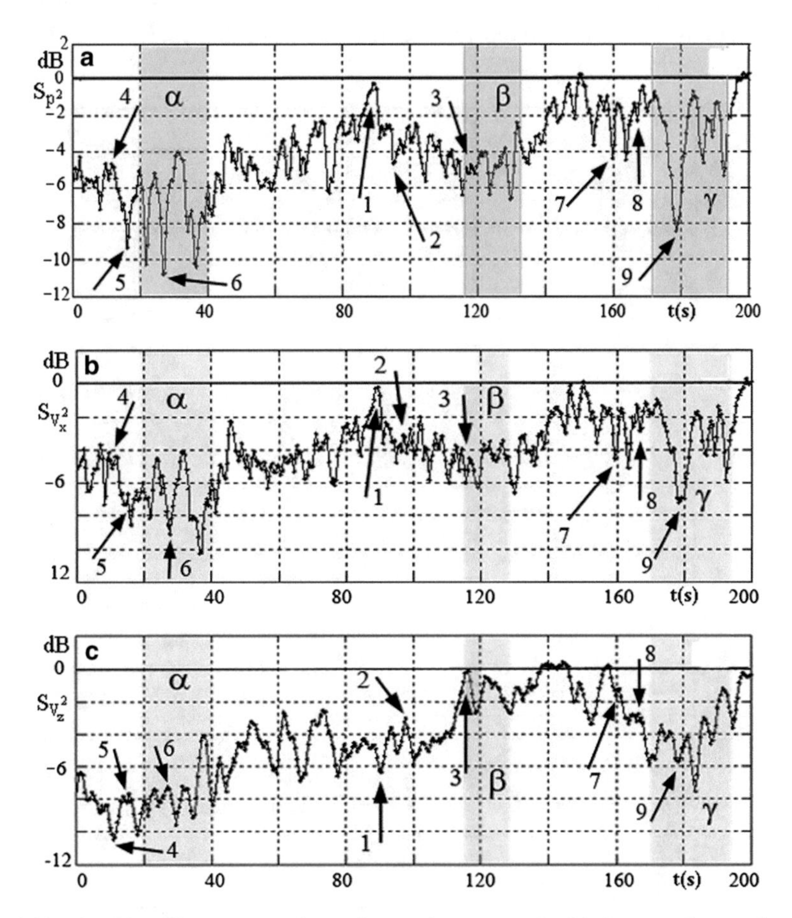

Fig. 4.11 Time dependence: **a** envelope of acoustic pressure $S_{p^2}(t)$; **b, c** envelopes of power $S_{V_x^2}(t)$ and $S_{V_z^2}(t)$, respectively. Averaging time is 1 s. Bandwidth: 88 Hz. The decibel scale is arbitrary

Envelopes of $p(t)$ and the x component of particle velocity $V_x(t)$ are similar, while the envelope of the z component of particle velocity $V_z(t)$ has a different dependence on time (Fig. 4.11). Interference minima of $p(t)$ and $V_x(t)$ at time intervals α, β, γ are associated with destructive interference (shaded). Level highs occur in areas of constructive interference (white background). Where interference is constructive, $\Delta\varphi_{pV_x}(t) = 0°$, $\mathrm{Re}\,\gamma_x(t) \approx +1.0$and $\mathrm{Im}\,\gamma_z(t) \approx 0$, meaning that $p(t)$ and $V_x(t)$ are in phase (Fig. 4.12). Conclusion: in the region of constructive interference, signal energy is carried along the waveguide axis by a coherent plane wave, as in the first experiment (Figs. 4.7 and 4.9), meaning that the condition of the mode theory is satisfied in this waveguide as well.

In Fig. 4.12, points 6, 3, 9 in the areas of destructive interference α, β, γ are the centres of local vortices [17, 18]. A local vortex is no more than 0.1λ, or ~ 1.7 m, in diameter. As shown in [17], in intervals α and γ there are vortices oscillating relative to the centre of the combined receiver, so fluctuations of $\Delta\varphi_{pV_x}$, $\mathrm{Re}\,\Gamma_x(t)$, $\mathrm{Im}\,\Gamma_z(t)$

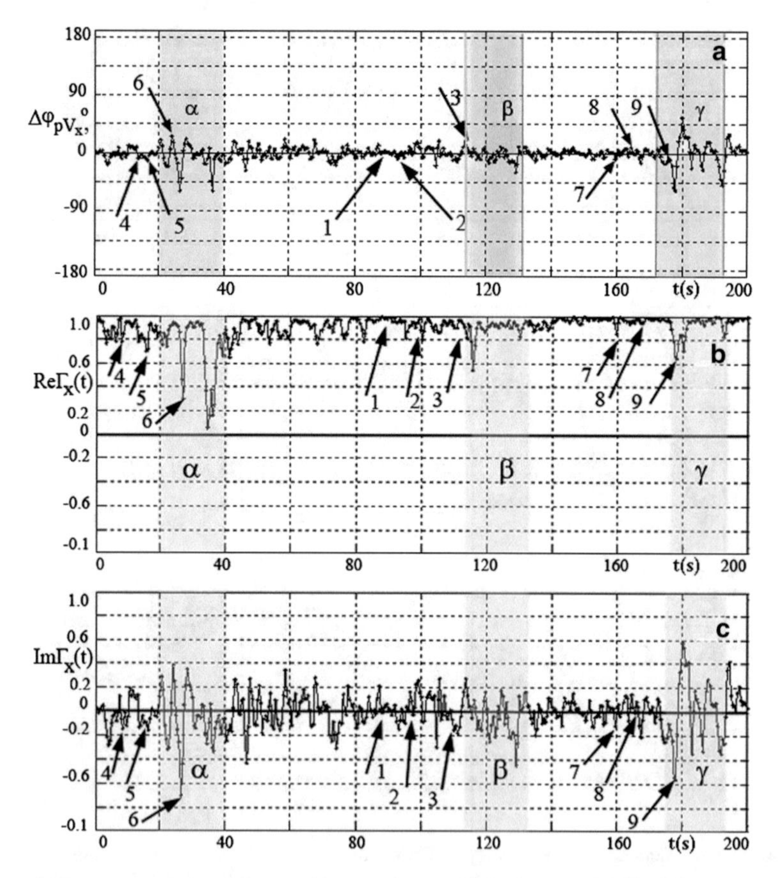

Fig. 4.12 Time dependence: **a** phase difference $\Delta\varphi_{pV_x}(t)$; **b** x component of the real part $\mathrm{Re}\,\Gamma_x(t)$; **c** x component of the imaginary part $\mathrm{Im}\,\Gamma_x(t)$. Averaging time is 1 s. Bandwidth: (88 ± 1) Hz

and significant 'dips' in $p(t)$ and $V_x(t)$ can be seen in Fig. 4.12 with an averaging time of 1 s. The averaging time of 1 s is long enough to see the microstructure of the local vortex in detail. In the β region there is a 'stationary' vortex, which can only be detected at averaging times ≤ 0.05 s. Vortices in the α, β, γ regions occupy the following time intervals: $\Delta T_\alpha = 12$ s, $\Delta T_\beta = 1$ s, $\Delta T_\gamma = 4$ s. Let us consider the z components of acoustic field: phase difference $\Delta\varphi_{pV_z}(t)$ and the z components of the real and imaginary parts of time coherence Re $\Gamma_z(t)$ and Im $\Gamma_z(t)$ (Fig. 4.13). In the first experiment, $\Delta\varphi_{pV_z}(t)$ fluctuated within $\pm\ \pi/4$ of $\pi/2$ (Figs. 4.8 and 4.10). In this experiment, $\Delta\varphi_{pV_z}(t)$ experiences a steady, time-linear advance of phase difference (Fig. 4.13a). It takes the phase difference $\Delta\varphi_{pV_z}(t)$ ~ 80 s to advance 2π. As a result, functions Re $\Gamma_z(t)$ and Im $\Gamma_z(t)$ have an alternating near-periodic structure in time (distance), since Re $\Gamma_z(t)$ ~ $\cos\Delta\varphi_{pV_z}(t)$, but Im $\Gamma_z(t)$ ~ $\sin\Delta\varphi_{pV_z}(t)$. The Im $\Gamma_z(t)$ curve is shifted with respect to Re $\Gamma_z(t)$ by a quarter of the period. With $-90° < \Delta\varphi_{pV_z}(t) < 90°$, the cosine is positive and Re $\Gamma_z(t) > 0$; at

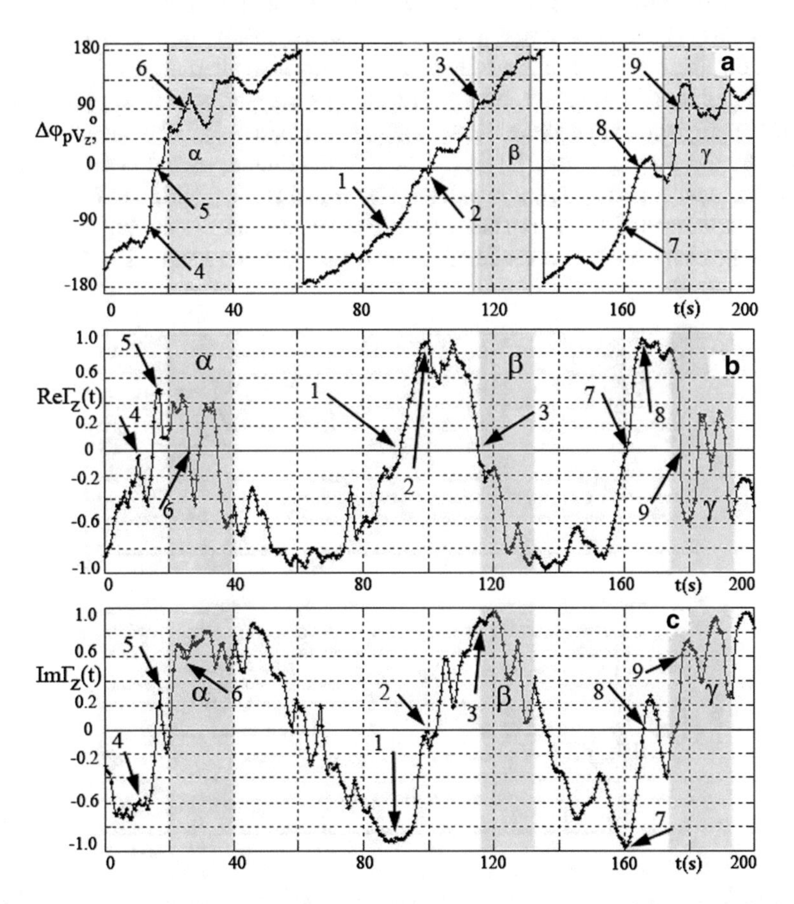

Fig. 4.13 Time dependence: **a** phase difference $\Delta\varphi_{pV_z}(t)$; **b** z component of the real part Re $\Gamma_z(t)$; **c** z component of the imaginary part Im $\Gamma_z(t)$. Averaging time: 1 s

angles 90° < $\Delta\varphi_{pV_z}(t)$ < – 90°, the cosine is negative and Re $\Gamma_z(t) < 0$. In Figs. 4.8 and 4.10 (first experiment) and Fig. 4.13 (second experiment), Re $\Gamma_z(t)$ behaves in a similar way. Re $\Gamma_z(t)$ is positive when signal energy travels from surface-to-bottom (in the + z direction) and negative when the direction is reversed. Thus, as in the first experiment (Figs. 4.7 and 4.10), there is an alternating flux of signal energy along the z *axis*. Let us consider the time interval from point 4 to point 7, which is two full periods with an advance of phase difference $\Delta\varphi_{pV_z}(t) = 4\pi$ (Fig. 4.13). Maxima of Re $\Gamma_z(t) = \pm 1.0$ correspond to $\Delta\varphi_{pV_z}(t) \approx 0°$ and $\Delta\varphi_{pV_z}(t) = \pm 180°$. Re $\Gamma_z(t)$ is zero at points 1, 3, 4, 6, 7, 9, where $\Delta\varphi_{pV_z}(t) = \pm \pi/2$ (Fig. 4.13a). Point 1 is where $S_{p^2}(t)$ has its maximum. At that point, $\Delta\varphi_{pV_z}(t) = -\pi/2$, and Re $\Gamma_z(t)$ passes through zero and changes sign from '–' to ' + '. Point 3 is where $S_{p^2}(t)$ has its minimum, with $\Delta\varphi_{pV_z}(t) = + \pi/2$ and Re $\Gamma_z(t)$ changing its sign from ' + ' to '–'. In the interims between points 4 and 6, 6 and 1, 1 and 3, 3 and 7, 7 and 9, Re $\Gamma_z(t)$ reaches a maximum of ± 1.0. This whole process completely matches the first experiment for Re $\Gamma_z(t)$, but Im $\Gamma_z(t)$ follows a different pattern.

Whereas in the first experiment Im $\Gamma_z(t)$ remained positive throughout (Figs. 4.8d and 4.10c), here Im $\Gamma_z(t)$ has a near-periodic sinusoidal alternating dependence on time. At points 1, 3, 4, 6, 7, 9, Im $\Gamma_z(t)$ reaches its extrema of ~ ± 1.0 (Fig. 4.13c). Signal energy streamline along the waveguide axis in the vertical plane $x0z$ is determined by the grazing angle $\theta(t)$ (Fig. 4.14). The $\pi/2$ jumps of $\theta(t)$ in the vicinity of the three local vortices α, β, γ match the theoretical curve (solid line) from [12].

The experimental $\theta(t)$ curve has a functional relationship (according to points 1–3) with the characteristics considered above (Figs. 4.11, 4.12 and 4.13). Component

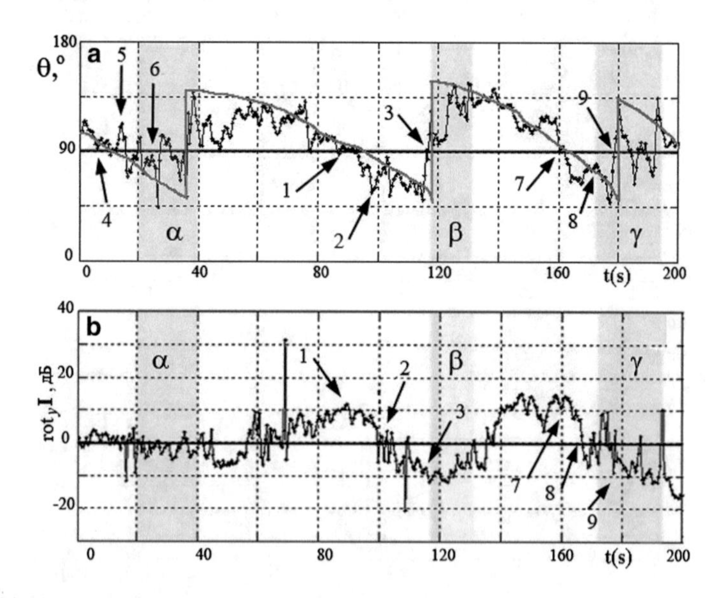

Fig. 4.14 Time dependence: **a** grazing angle of energy flow $\theta(t)$; **b** y component *of* rotation $\mathbf{rot}_y\boldsymbol{I}(t)$. Frequency: 88 Hz, averaging time: 1 s. $\theta = 90°$ describes the x axis

of the intensity vector $\boldsymbol{I}_{xz} = \mathbf{i}I_x(t) + \mathbf{k}I_z(t)$, which is tangential to the $\theta(t)$ curve, also deviates from the x axis in the vicinity of local vortices, which causes the intensity vector field to swirl. As acoustic signal energy flows around the vortices, vorticity develops in the energy flux density vector, with a nonzero z component of the energy flux. This way, the presence of the vortex structure 'bends' the streamlines of acoustic energy, creating eddies in the energy flow with nonzero horizontal components of rotation of the intensity vector [6, 7, 18]. The degree of vorticity in this case is determined by the y component of rotation of the intensity vector $\mathbf{rot}_y\,\boldsymbol{I}(t)$ (Fig. 4.14b). All these characteristics of the field are deterministic functions and are interconnected. The sign of the rotation is opposite to the sign of Im $\Gamma_z(t)$ [see (4.6)]. The rotation is positive where energy moves anticlockwise (polarisation of particle motion is positive) and negative where energy moves clockwise (polarisation of particle motion is negative). Let us consider the relationship among the functions (Figs. 4.11, 4.12, 4.13 and 4.14). At point 1, Re $\Gamma_z(t) = 0$, Im $\Gamma_z(t) = -1.0$, $\mathbf{rot}_y\,\boldsymbol{I}(t) > 0$. At point 2, Re $\Gamma_z(t) = +1.0$, Im $\Gamma_z(t) = 0$, $\mathbf{rot}_y\,\boldsymbol{I}(t) > 0$. At point 3, Re $\Gamma_z(t) = 0$, Im $\Gamma_z(t) = +1.0$, $\mathbf{rot}_y\,\boldsymbol{I}(t) > 0$.

At point 7, Re $\Gamma_z(t) = 0$, Im $\Gamma_z(t) = -1.0$, $\mathbf{rot}_y\,\boldsymbol{I}(t) > 0$. At point 8, Re $\Gamma_z(t) = +1.0$, Im $\Gamma_z(t) = 0$, $\mathbf{rot}_y\,\boldsymbol{I}(t) = 0$.

In other words, energy in the eddy field 'pours' through a chain of vortices of opposite signs. The rotation has its maxima where the vector $\boldsymbol{I}_{xz}(t)$ intersects the waveguide axis (points 1, 3, 7, 9); there, Re $\Gamma_z(t) = 0$ and Im $\Gamma_z(t) = \pm 1.0$. Note that the $\theta(t)$ curves are similar between the first and the second experiment (Figs. 4.10d and Fig. 4.14a). Hence it follows that in the first experiment, local vortices must exist at the depths adjacent to the receiver, meaning that vorticity exists throughout the waveguide water column.

4.3.2.3 Statistical Analysis of the Data from the First and Second Experiments

Our analysis of experimental data using deterministic functions (Sects. 4.1.2 and 4.2.1) yields a deterministic mechanism of signal energy transfer in the shallow water waveguide. The experimental data are time series of random functions $p(t)$, $V_x(t)$, $V_y(t)$, $V_z(t)$. In our statistical analysis we consider the random acoustic field to be stationary and ergodic, and the functions of interest to be independent and Gaussian [3].

Consider now probability density histograms $P(\varphi_y)$, $P(\mathrm{Re}\,\gamma_y)$, $P(\mathrm{Im}\,\gamma_y)$, $P(\varphi_z)$, $P(\mathrm{Re}\,\gamma_z)$, $P(\mathrm{Im}\,\gamma_z)$ of the following functions in the first experiment: $\Delta\varphi_{pV_j}(t)$, $\mathrm{Re}\,\gamma_j(t)$, $\mathrm{Im}\,\gamma_j(t)$, where $j = x, y, z$. Figures 4.15 and 4.16 show histograms of the movement of the 163 Hz tone energy. The pattern of movement in the horizontal plane indicates a high probability of observing horizontal flow of signal energy: for $\Delta\varphi_{pV_y}(t)$, around (0.8–0.9) in the ($0° \pm 8°$) range; for Re $\Gamma_y(t) \approx +1.0$ the probability is (0.8–0.9); Im $\Gamma_y(t) \approx 0$ with a probability of ~ (0.7–0.8) (Fig. 4.15).

Statistical analysis shows that in the vertical plane along the z axis, a standing wavefield and an alternating energy flux are equally possible—in other words, both

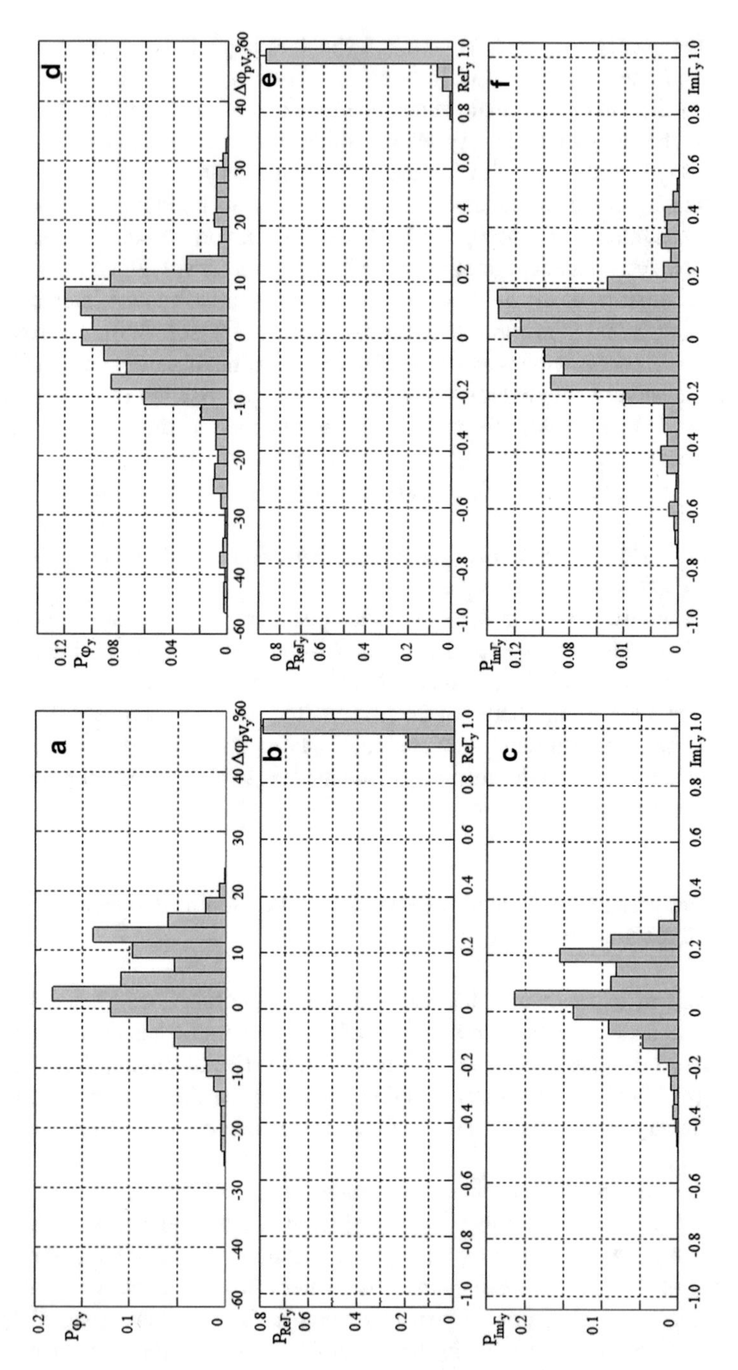

Fig. 4.15 Probability density histograms of horizontal characteristics of the acoustic field. **a–c** 9050–7250 m distance; **d–f** 3650–1850 m. **a, d** histograms of $\Delta\varphi_{pV_y}(t)$; **b, e** Re $\Gamma_z(t)$; **c, f** Im $\Gamma_z(t)$. Averaging time is 1 s. Accumulation time: 1700 s. Tone frequency: 163 Hz

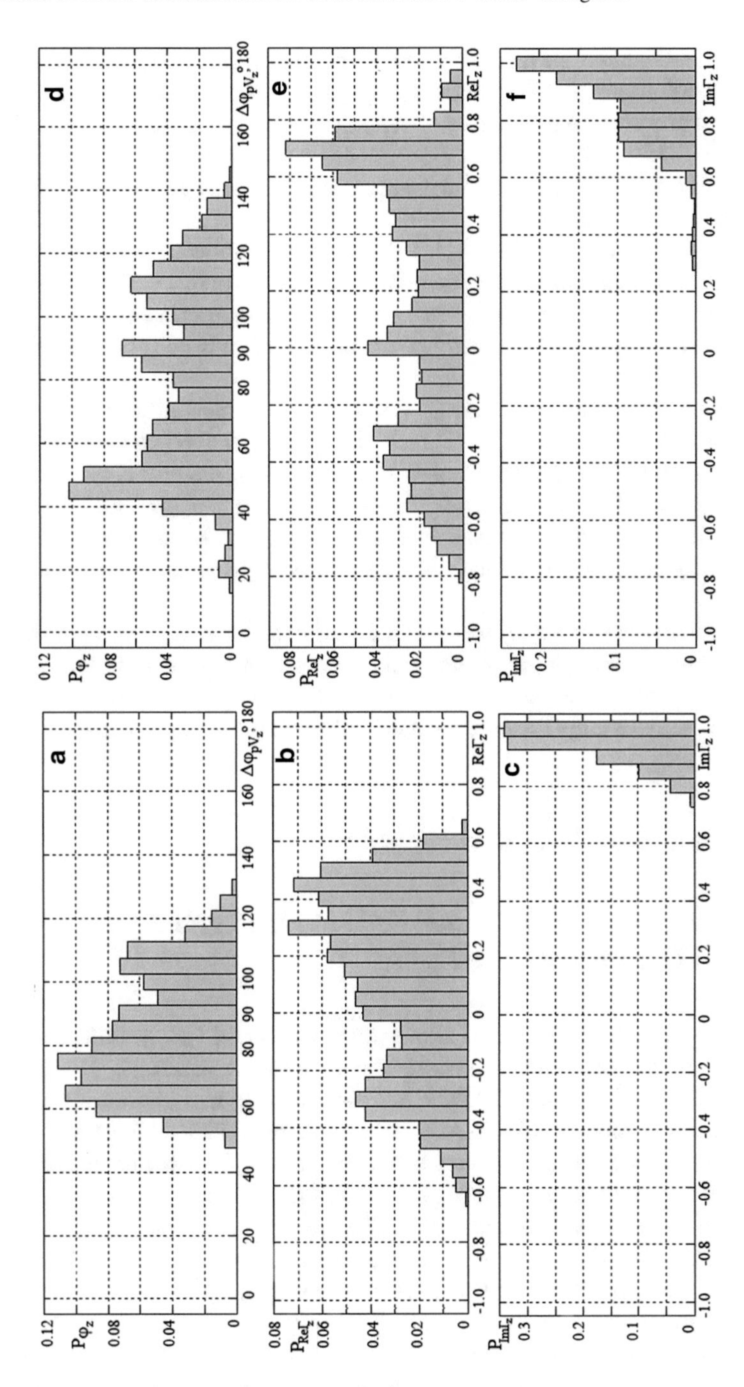

Fig. 4.16 Probability density histograms of vertical characteristics of the acoustic field. **a–c** 9050–7250 m distance; **d–f** 3650–1850 m. **a**, **d** histograms of $\Delta\varphi_{pV_z}(t)$; **b**, **e** Re $\Gamma_z(t)$; **c**, **f** Im $\Gamma_z(t)$. Averaging time is 1 s. Accumulation time: 1700 s. Tone frequency: 163 Hz

travelling and standing waves are observed to overlap in the vertical plane. Probability density distributions are asymmetrical (Fig. 4.16). Upward flows of energy (surface-to-bottom) prevail. The probability of $\Delta\varphi_{pV_z}(t)$ in the $(90 \pm 10°)$ region is no more than (0.23–0.35); with $\mathrm{Im}\,\Gamma_z(t) \approx +1.0$, it can be as high as ~ 0.7; $\mathrm{Re}\,\Gamma_z(t)$ occupies the interval from − 0.8 to + 0.8, and the probability of $\mathrm{Re}\,\Gamma_z(t) = +(0.0\text{–}0.8)$ ranges from 0.6 to 0.7 as a result.

According to the mode theory [11, 19], the vertical flow of signal energy in the ideal Pekeris waveguide must be zero. The experiment points to the opposite, even though the chosen real waveguide was deemed to be a uniform one.

Statistical analysis of the second experiment (88 Hz) is consistent with the first one (Figs. 4.17 and 4.18).

Horizontal phase difference $\Delta\varphi_{pV_x}(t) \approx 0°$ with a probability of ~ 0.9; $\mathrm{Re}\,\Gamma_x(t)$ is in the + (0.9–1.0) range and $\mathrm{Im}\,\Gamma_x(t)$ in the $+(0.0 \pm 0.2)$ range with a probability of 0.8 all indicate the movement of tone signal energy along the waveguide axis. The histograms of the vertical components of the field $\Delta\varphi_{pV_z}(t)$ and $\mathrm{Re}\,\Gamma_z(t)$ are asymmetrical; the vertical bottom-to-surface flow prevails over the surface-to-bottom flow; density peaks at – 0.4 and + 0.7 are ~ 0.05; distribution of $\mathrm{Im}\,\Gamma_z(\mathrm{t})$ is near-uniform.

In summary, the first and the second experiment are consistent with the mode theory in the way energy propagates along the horizontal axis of the shallow water waveguide. In the transverse direction, along with standing waves, we have detected alternating signal energy fluxes associated with energy swirling due to local vortices.

4.4 Dynamics of Local Vortices

4.4.1 Properties of the Vector Field in the Region of Destructive Interference

Individual vortices in the real shallow water waveguide are identified using the properties of the deterministic theoretical model of a vortex [7, 10]. In statistical processing of experimental data, isolating a vortex in a given destructive area consisted in the following [18]. At averaging times (0.025–0.05), we found time intervals in the regions of destructive interference α, β, γ that contained relative minima of the envelope of acoustic pressure $p(t)$ (Fig. 4.11).

Next, in these intervals we calculated differential phase characteristics $\Delta\varphi_{pV_j}(t)$, components of the time coherence function $\mathrm{Re}\,\Gamma_j(t)$ and $\mathrm{Im}\,\Gamma_j(t)$, components of rotation of the intensity vector $\mathbf{rot}\,I_j(t)$ (where $j = x, y, z$) and grazing angle $\theta(t)$ of the energy streamline [see (4.12), (4.21)–(4.23)]. The resulting functions of time are independent characteristics of the acoustic vector field and make up a complete and consistent system of functions describing this phenomenon. 'Relative minima' were used because the acoustic field is a mixture of signal and underwater ambient noise, and an acoustic pressure minimum of the signal interference field will be constrained

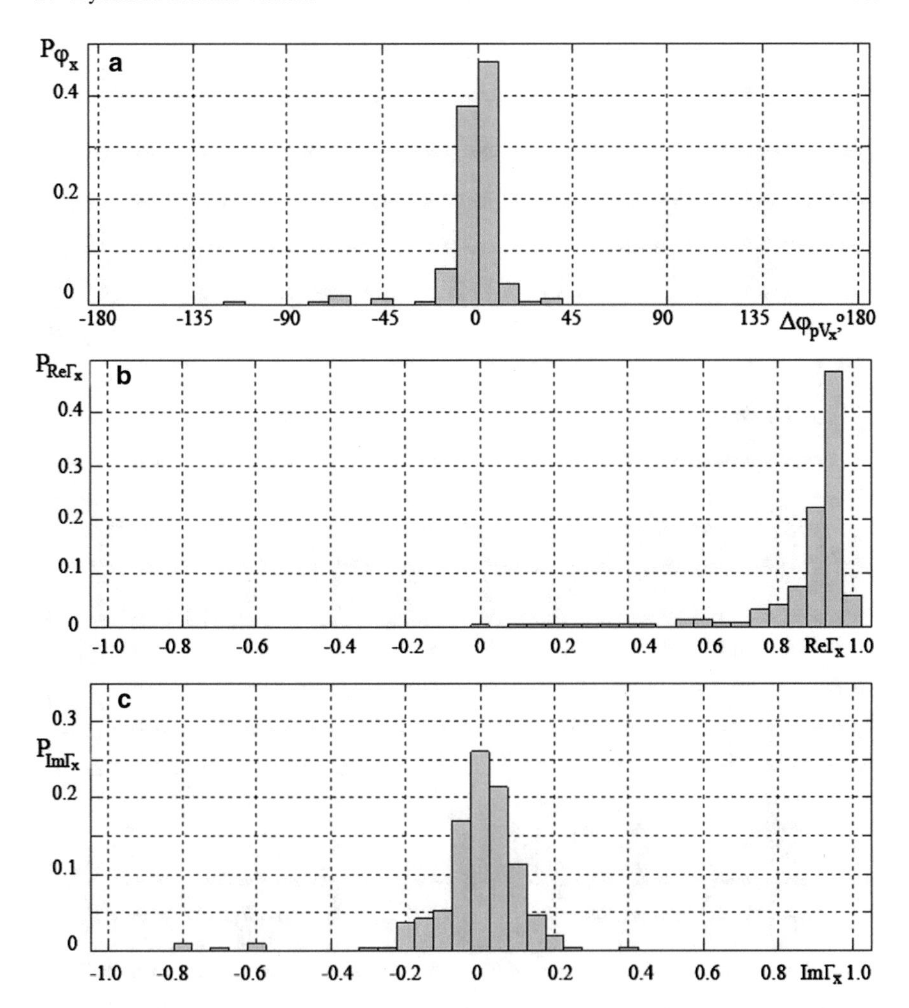

Fig. 4.17 Probability density histograms of horizontal characteristics of the acoustic field with three local vortices in the waveguide. Histograms: **a** $\Delta\varphi_{pV_x}(t)$, **b** Re $\Gamma_x(t)$, **c** Im $\Gamma_x(t)$. Averaging time is 1 s. Accumulation time: 200 s. Tone frequency: 88 Hz

by the signal-to-noise ratio. The top inset in Fig. 4.6 shows the layout of receivers 1–4. We use experimental data from receivers 1 and 2.

Let us consider the area of destructive interference α in which a physical object was detected that fits the description of a vortex.

Vortex $\boldsymbol{\alpha}$ (Figs. 4.19, 4.20 and 4.21). Vortex α is observed in the time realisation interval $\Delta T = 26 - 38$ s $= 12$ s. The chosen averaging time was $\Delta t = 0.025$ s, equivalent to a spatial averaging interval of ~ 0.04 m; this allows the ~ 1.7 m vortex to be examined in detail. The time interval of 12 s is equivalent to a spatial interval of 18 m, comparable with the wavelength $\lambda = 17$ m. The receiver-to-source distance is ~ 700 m. Pressure envelope (Fig. 4.19a) exhibits two regions

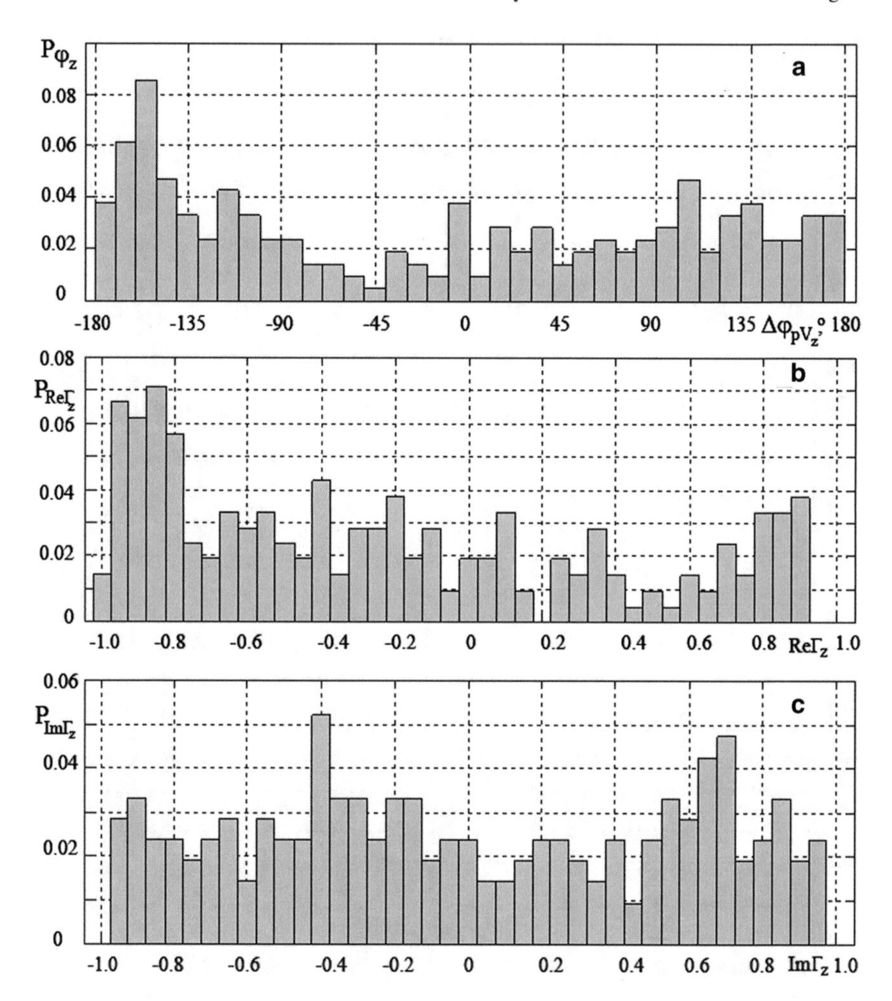

Fig. 4.18 Probability density histograms of vertical characteristics of the acoustic field with three local vortices in the waveguide. Histograms: **a** $\Delta\varphi_{pV_z}(t)$, **b** Re $\Gamma_z(t)$, **c** Im $\Gamma_z(t)$. Averaging time: 1 s. Accumulation time: 200 s. Tone frequency: 88 Hz

of relative minima. The first one is *ab* (~ 1.8 s long), the second *cd* (~ 5 s long). The envelope of $V_x(t)$ is similar to the envelope of $p(t)$. The envelope of $V_z(t)$ follows a different pattern (Fig. 4.19b). The relative minima of the envelopes can be significant (~ 12 dB), consistent with the signal-to-noise ratio in the α region with an averaging time of 0.025 s. The time interval *bc* between the relative minima of $p(t)$ is ~ 5 s long, or ~ 7 m in distance terms. Pressure in the time interval *bc* (Fig. 4.19a, point 2) equals the pressure in the area of constructive interference (marked with an asterisk * in Figs. 4.19, 4.20 and 4.21).

As is known [10], the envelope of $p(t)$ must have its minimum near the centre of the vortex and its maxima near the saddle; the minima of particle velocity must be near

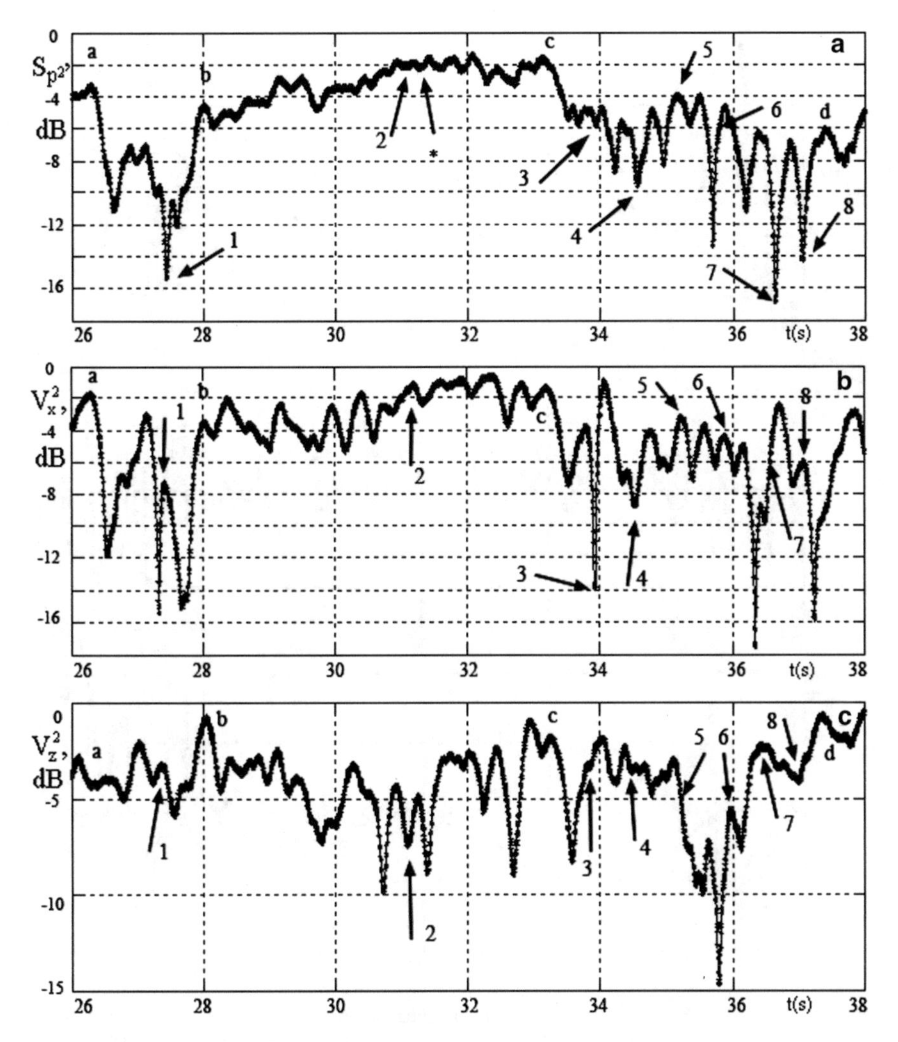

Fig. 4.19 Vortex α. First receiver. Envelopes of spectral power density as a function of time: **a** acoustic pressure $S_{p^2}(t)$; **b** x component of particle velocity $S_{V_x^2}(t)$; **c** z component of particle velocity $S_{V_z^2}(t)$. Frequency: 88 Hz. Averaging time: 0.025 s. The decibel scale is arbitrary

the saddle, but its maximum must be near the centre of the vortex. The minima of $p(t)$, $V_x(t)$, $V_z(t)$ occur at different times, as well they should (Fig. 4.19). All the functions of interest have anomalies in the time intervals *ab* and *cd* (Figs. 4.19, 4.20 and 4.21). Based on the criteria from [7, 10, 12], we identify the interference structures in the intervals *ab* and *cd* as vortices of the intensity vector. Consider the interval *ab*. Note the following feature in the vicinity of point 1: $p(t)$ has a minimum of ~ – 12 dB; phase difference $\Delta\varphi_{pV_z}(t)$ experiences a jump of π, resulting in Re $\Gamma_z(t)$ passing through zero and changing sign from ‘ + ’ to ‘–’ (the vector rotates anticlockwise);

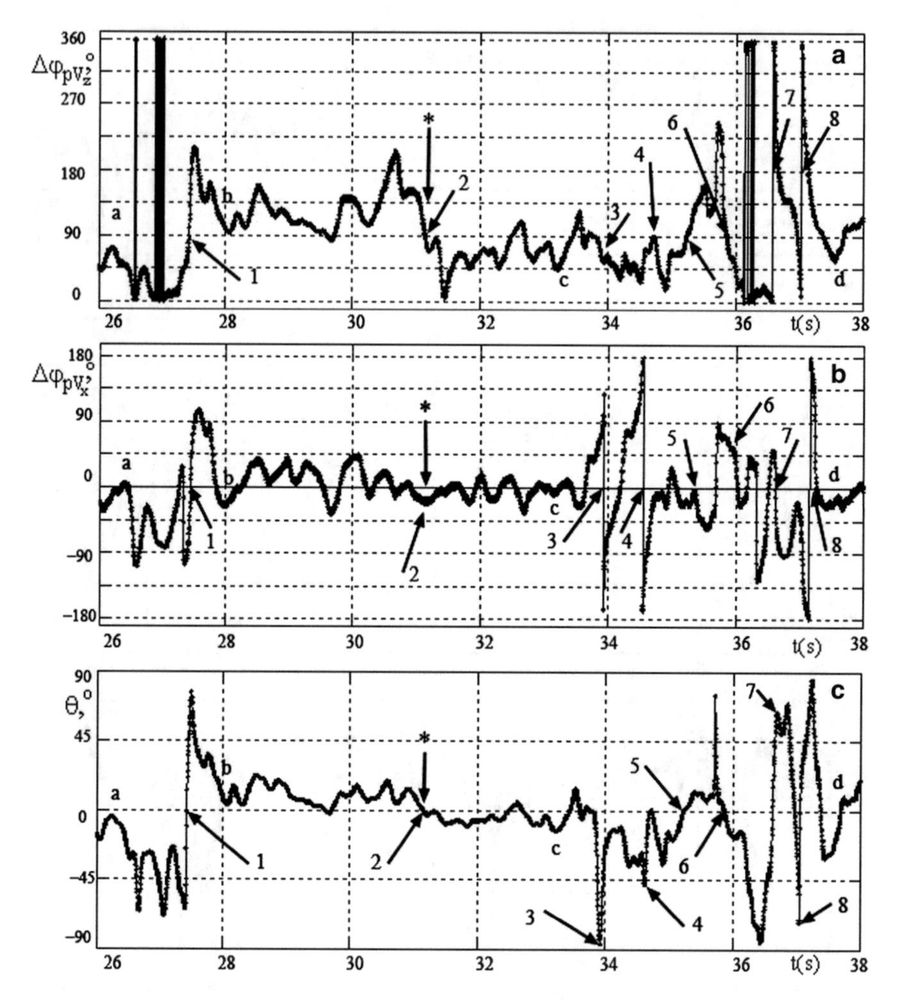

Fig. 4.20 Vortex α. First receiver. Time dependence: **a** phase difference $\Delta\varphi_{pV_z}(t)$; **b** phase difference $\Delta\varphi_{pV_x}(t)$; ***c*** grazing angle $\theta(t)$ of the energy streamline. Averaging time: 0.025 s

Im $\Gamma_z(t)$ experiences a jump and passes through zero; Re $\Gamma_x(t)$ jumps from + 1.0 to zero; Im $\Gamma_x(t)$ jumps and passes through zero; θ passes through 0° and experiences a jump > $\pi/2$.

The reciprocal synchronous dynamics of these parameters indicates that the phase centre of the combined receiver is near the centre of a vortex (point 1). The totality of the observed features exactly fits the model of a ' + ' vortex [7, 10]. However, the following departure from the ideal vortex model is observed. Re $\Gamma_x(t)$ has four minima, but according to the criterion for an ideal vortex there must be only one minimum at point 1 (Fig. 4.21c).

This is because phase difference $\Delta\varphi_{pV_x}(t)$ makes a sequence of oscillations in the interval *ab* within 0° → –90° → 0° → –90° → 0° → –90° → 0° → + 90°

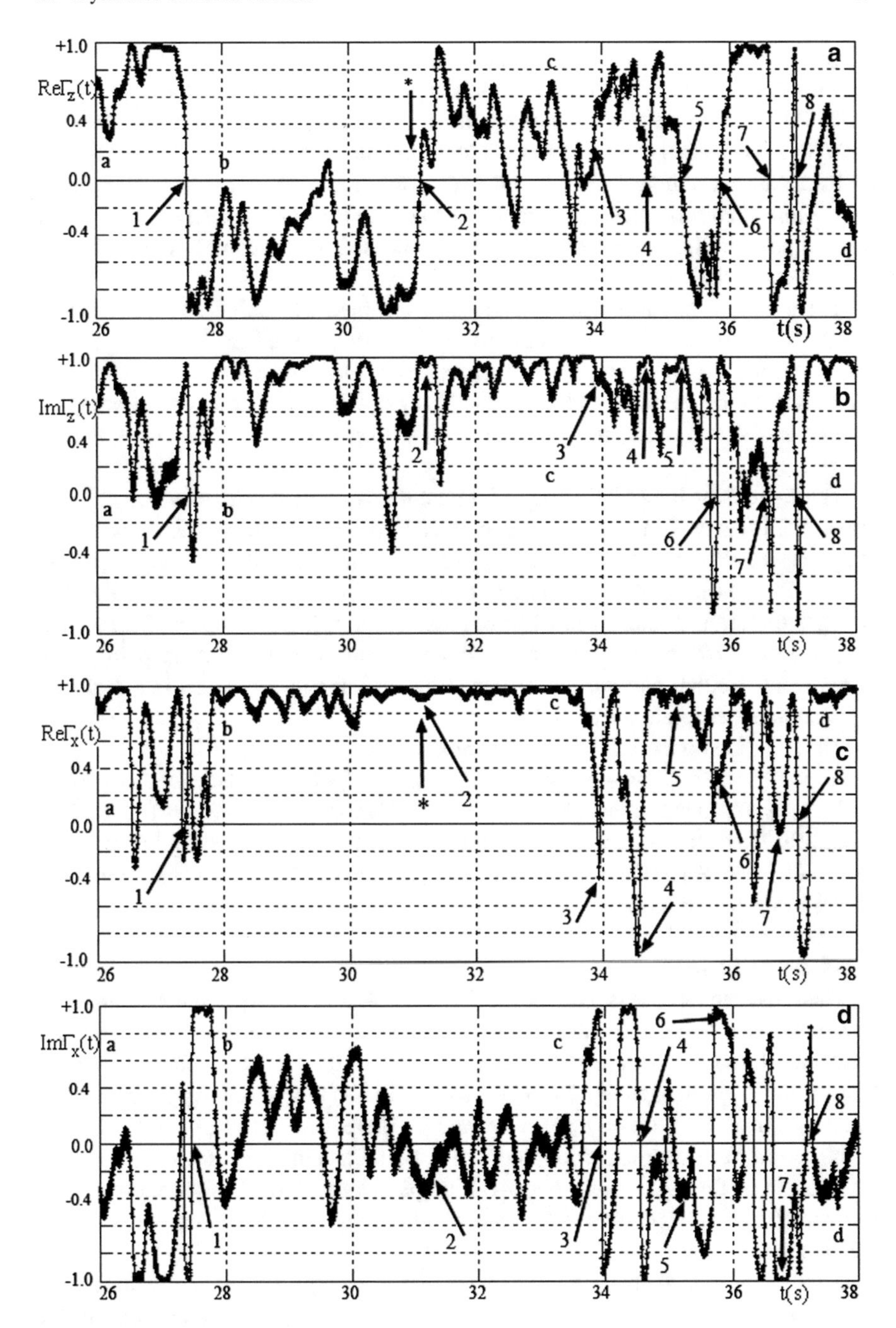

Fig. 4.21 Vortex α. First receiver. Time dependence: **a** real part of the z component of the time coherence function Re $\Gamma_z(t)$; **b** its imaginary part Im $\Gamma_z(t)$; **c** Re $\Gamma_x(t)$; **d** Im $\Gamma_x(t)$. Averaging time: 0.025 s

(Fig. 4.20b), causing Re $\Gamma_x(t)$ to periodically vary between + 1.0 and 0 and on to – 0.4 (Fig. 4.21c). The total advance of the phase difference $\Delta\varphi_{pV_x}(t)$ is 4π. Since Re $\Gamma_z(t)$ doesn't experience significant fluctuations in the process (the only time Re $\Gamma_z(t)$ passes through zero is not a random fluctuation), it can be assumed that the vortex oscillates relative to the receiver within the segment *ab* along the *x* axis. More precisely, these are oscillations (instability) of the interference field at a given point in the real waveguide, possibly caused by variation of hydrophysical parameters of the medium.

To summarise, we have been able to record an object that satisfies the criteria for a vortex of the acoustic intensity vector in the stationary case and an ideal waveguide [7, 10] but which also possesses a new the property associated with its movement along the *x* axis. Since the phase advance $\Delta\varphi_{pV_x}(t)$ equals 4π, the number of complete oscillations is two, and therefore, the displacement frequency is 1 Hz in the 2 s interval.

Throughout the interval *bc*: $\Delta\varphi_{pV_x}(t)$ fluctuates near zero (Fig. 4.20b); a coherent signal field sets in along the + *x* axis and Re $\Gamma_x(t) \approx + 1.0$ (Fig. 4.21c); at point 2 $\Delta\varphi_{pV_z}(t) = 90°$, $\psi(t) = 0°$ and Re $\Gamma_z(t)$ passes through zero and changes sign from '–' to ' + ', meaning that the vertical component of the intensity vector Re $\Gamma_z(t)$ changes direction through 180°. But because at point 2 $\Delta\varphi_{pV_z}(t) = 90°$, Im $\Gamma_z(t)$ $\approx + 1.0$, Re $\Gamma_x(t) \approx + 1.0$, it follows that this point is not a singular point of the phase front. Neither can vortices occur in the *bc* interval, since this is a region of constructive interference (Fig. 4.19a).

The situation on the *cd* interval is similar to *ab*.

At point 3: *p*(*t*) fell by ~ 3 dB; $S_{V_x^2}$ lost ~ 12 dB; $S_{V_z^2}$ hasn't changed; $\Delta\varphi_{pV_z}(t) \approx \pi/2$; $\Delta\varphi_{pV_x}(t)$ makes a leap $\geq \pi$; $\theta(t)$ makes a leap $0° \to -\pi/2 \to 0°$; Re $\Gamma_z(t)$ ≈ 0.2; Re $\Gamma_x(t) \approx - 0.4$; Im $\Gamma_z(t) + 0.8$; Im $\Gamma_x(t)$ jump-changes sign from + 1.0 to – 1.0. Because Re $\Gamma_x(t) < 0$, that is, signal energy flows towards the source and $S_{V_x^2}(t)$ has a deep minimum.

From which it follows that point 3 is between the centre and the saddle of a vortex, but closer to the saddle.

Then, the same process repeats itself at point 4, that is, displacement causes the vortex to leave point 3, then the vortex returns there, point 3 and point 4—this is recording the same position of the vortex relative to the receiver's phase centre. At point 5, Re $\Gamma_z(t)$ changes sign from ' + ' to '–'; Re $\Gamma_x(t) \to + 1.0$, $\Delta\varphi_{pV_z}(t) = \pi/2$; but $\Delta\varphi_{pV_x}(t) = 0°$; $\psi(t) = 0°$; Im $\Gamma_z(t) \pm 1.0$. There is no vortex at point 5. Point 6 is the centre of the vortex, since Re $\Gamma_z(t) = 0$ and changes sign from '–' to ' + '; Im $\Gamma_z(t) = 0$; Re $\Gamma_x(t) \to 0$; $\Delta\varphi_{pV_x}(t)$ jumps from $\pi/2$ to 0°. Next, the centre of the vortex moves to points 7 and 8 before returning to point 6. The displacement frequency is ~ 2 Hz.

In the *cd* interval, there is a vortex with a '–' sign. Energy flux circulates clockwise. With $\Delta\varphi_{pV_x}(t)$ changing regularly, phase difference $\Delta\varphi_{pV_z}(t)$ varies little in the 45°–90° range. This means that regular vortex displacements along the *x* axis are also observed in the interval *cd*.

Vortex β is observed in the destructive interference region β in the time interval $\Delta T = 118\text{ s} - 128\text{ s} = 10\text{ s}$, equivalent to a spatial interval of 15 m. The receiver and

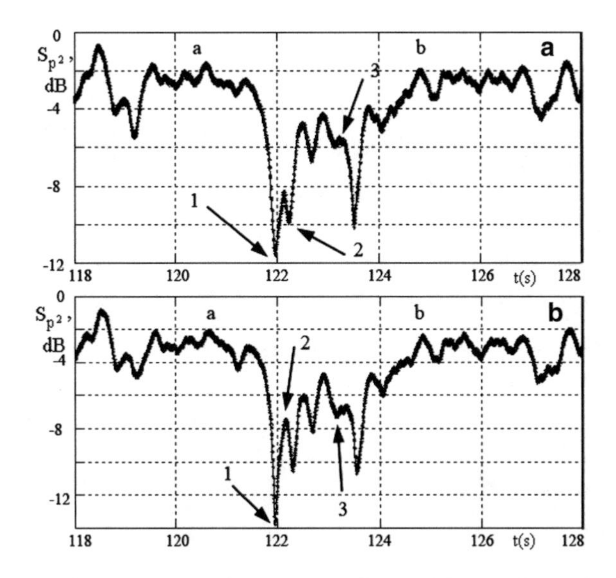

Fig. 4.22 Vortex β. Time dependence of the envelope of spectral density of acoustic pressure $S_{p^2}(t)$. **a** First receiver; **b** second receiver. Frequency: 88 Hz. Averaging time is 0.025 s. The decibel scale is arbitrary

the source are ~ 520 m apart. Averaging time is 0.025 s. Spatial averaging interval is ~ 0.04 m. Figures 4.22, 4.23 and 4.24 plot these functions: $S_{p^2}(t)$, $\Delta\varphi_{pV_z}(t)$, $\Delta\varphi_{pV_x}(t)$, Re $\gamma_x(t)$. Figures 4.22, 4.23 and 4.24 show that the combined receiver is in the region of a vortex whose structure is different to that of the α vortex.

It must be noted that the observed vortex structures depend primarily on which region of the vortex the receiver's acoustic centre is in. Either way, the fundamental distinctive features of the vortex are detected.

Receivers 1 and 2 are spaced 0.67 m vertically (Fig. 4.6). The 'dips' of pressure levels in the interval *ab* at point 1 are: ~ 10 dB at the first receiver, ~ 12 dB at the second receiver. This means that the second receiver is closer to the centre than the first. The general appearance of the curves $p(t)$ is similar (Fig. 4.22). From Figs. 4.23 and 4.24, it follows that between 118 and 122 s phase difference $\Delta\varphi_{pV_z}(t)$ oscillates about 180° while $\Delta\varphi_{pV_x}(t)$ fluctuates about 0°. In this interval, Re $\Gamma_z(t) \approx -1.0$, meaning that energy flows from the bottom to the surface (Re $\Gamma_z(t)$ is not charted here), and Re $\Gamma_x(t) = +1.0$, hence energy flows in the $+x$ direction. In this time interval, both receivers are in the region of constructive interference (Figs. 4.22 and 4.23)—in other words, outside the vortex.

All the functions behave anomalously in the vicinity of $t = 122$ s in Figs. 4.22, 4.23 and 4.24 meaning that the acoustic centres of the receivers have entered the vortex region. The functions of the first and the second receiver behave differently at $t > 122$. Realisation at the first receiver at $t = 122$ s (Fig. 4.23): $\Delta\varphi_{pV_z}(t)$ experiences a jump of π; $\Delta\varphi_{pV_x}(t)$ jumps 2π; Re $\Gamma_x(t)$ jumps from $+1.0$ to -1.0 in the span of ~ 0.4 s.

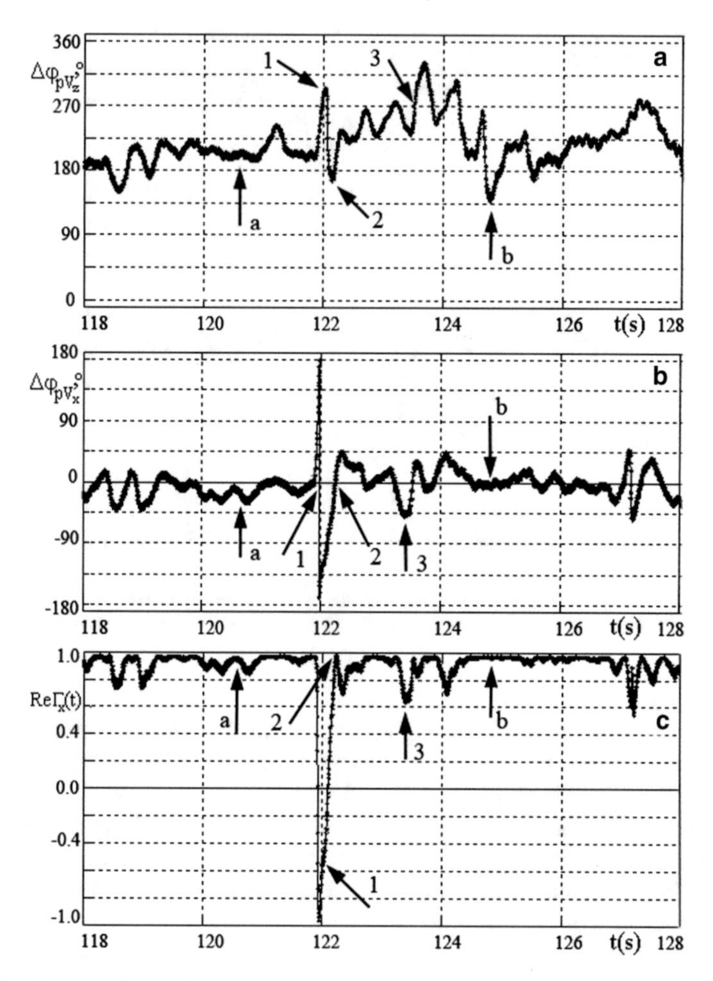

Fig. 4.23 Vortex β. First receiver. Time dependence: **a** phase difference $\Delta\varphi_{pV_z}(t)$; **b** phase difference $\Delta\varphi_{pV_x}(t)$; **c** real part of the x component of time coherence function Re $\Gamma_x(t)$. Averaging time: 0.025 s

Therefore, at $t = 122$ s Re $\Gamma_x(t)$ records the x component of the signal intensity vector pointed towards the signal source, meaning that the acoustic centre of the first receiver is in an acoustic vortex between the centre and the saddle. Further on, from $t = 122$ s to point b, the influence of the vortex is insignificant as the receiver enters the region of constructive interference. As the receiver passes through the vortex, Re $\Gamma_z(t)$ changes from -1.0 to $\sim +1.0$, meaning there is a change of sign from '–' to ' + ', hence the vortex has a negative sign. The second receiver in the time interval *ab* passes through a vortex twice, at points 1 and 3 (Fig. 4.24); the sign of the vortex is positive, since Re $\Gamma_z(t)$ changes sign from ' + ' to '–'. In this case, we have two closely spaced vertical vortices in the region β.

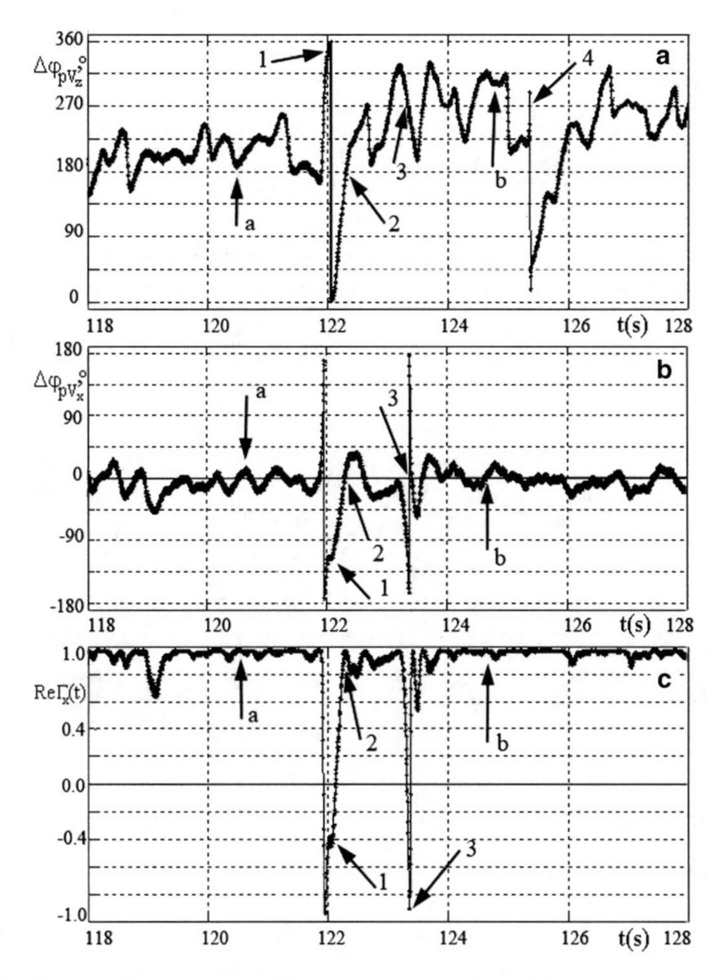

Fig. 4.24 Vortex β. Second receiver. Time dependence: **a** phase difference $\Delta\varphi_{pV_z}(t)$; **b** phase difference $\Delta\varphi_{pV_x}(t)$; **c** real part of the x component of time coherence function Re $\Gamma_x(t)$. Averaging time: 0.025 s

Vortex γ. Vortex γ is observed in the region of destructive interference γ in the time interval $\Delta T = 175 - 183$ s $= 8$ s. The vortex proper occupies the interval from 178 to 181 s, or ~ 3 s, equivalent to a distance of ~ 4.5 m (Fig. 4.25). The source and the receiver are ~ 400 m apart.

Time intervals from 175 s to point a and from point b to 183 s are in the region of constructive interference. In these intervals, Re $\Gamma_x(t) \approx +1.0$ and $\Delta\varphi_{pV_x}(t) \approx 0°$, hence the x component of intensity points in the $+x$ direction. However, in the interval ab at points 4, 6, 10, 14 Re $\Gamma_x(t) = +1.0$ and $\Delta\varphi_{pV_x}(t) = 0°$—that is, the receiver intermittently returns to the region of constructive interference. It follows that the interference field experiences a time shift along the x axis relative to the phase centre of the receiver, which is stationary. Figure 4.25 depicts three identical fluctuations

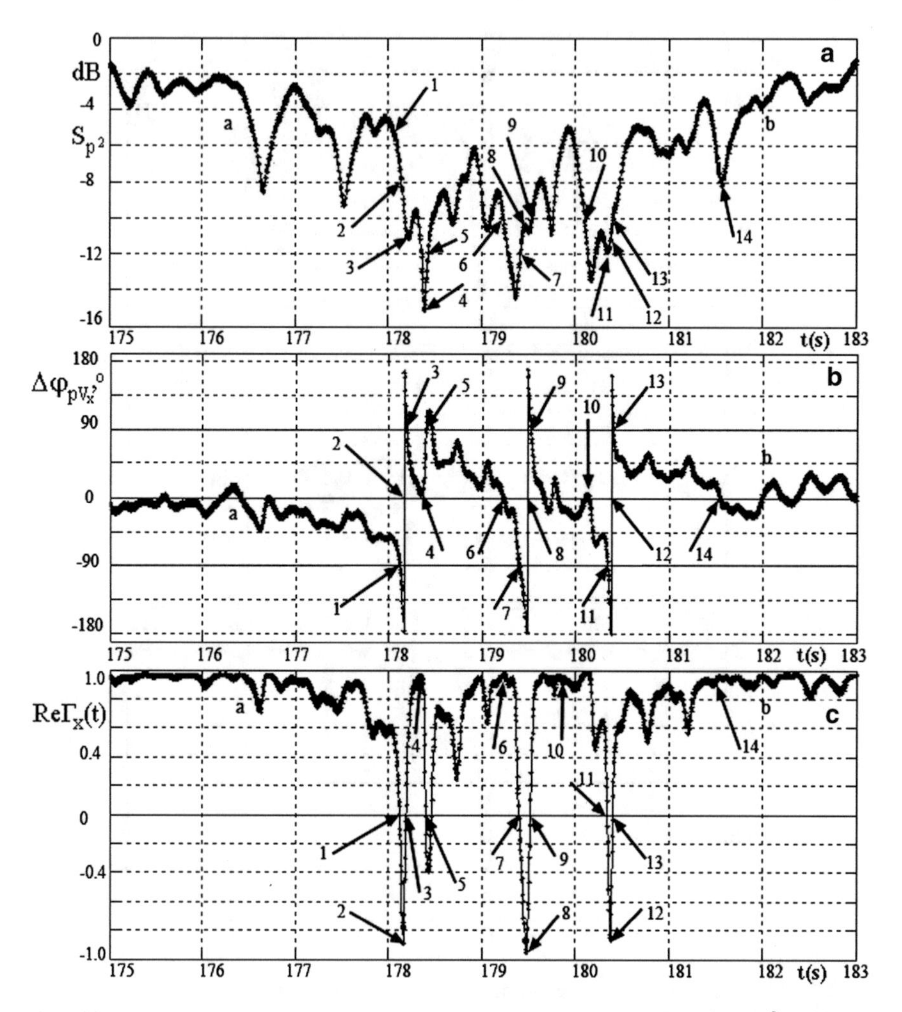

Fig. 4.25 Vortex γ. First receiver. Time dependence: **a** envelope of acoustic pressure $p^2(t)$; **b** phase difference $\Delta\varphi_{pV_x}(t)$; **c** real part of the x component of time coherence function Re $\Gamma_x(t)$. Averaging time: 0.025 s

of $p(t)$, $\Delta\varphi_{pV_x}(t)$ and Re $\Gamma_x(t)$. From point a to point 4: $p(t)$ fluctuates and decays, its level dropping by ~ 13 dB at point 4 (area of destructive interference); $\Delta\varphi_{pV_x}(t)$ rotates through 2π clockwise ($0° \to \pi/2 \to -\pi \to +\pi/2 \to 0°$);); Re $\Gamma_x(t)$ reaches ~ + 1.0 at points a and 4, but at point 2 Re $\Gamma_x(t) \approx -1.0$, which indicates the vortex region between the centre and the saddle. Energy at point 2 flows towards the source.

Next the oscillating process returns to its original state: while at point 1 $\Delta\varphi_{pV_x}(t) = -(\pi/2)$ and then tends to $-\pi$, in the interval 4–6 $\Delta\varphi_{pV_x}(t)$ varies within $0° \to +\pi/2 \to 0°$. From point 6 to point 10, the second fluctuation repeats the first. Re $\Gamma_x(t)$ behaves similarly to the first instance from point 6 to point 10, with Re $\Gamma_x(t) = -1.0$ at point 8. The third fluctuation of $\Delta\varphi_{pV_x}(t)$ and Re $\Gamma_x(t)$ from point 10 to point 14 repeats the first and the second. The receiver is between the centre and the saddle. The third fluctuation is completely identical to the first and the second. The total advance of phase difference of Re $\Gamma_x(t)$ is 6π; that is, vortex displacement frequency is 1 Hz.

Vortex δ. Turning now to the vortex observed a considerable ~ 1000 m away from the sound source. Radiation frequency is 88 Hz (Fig. 4.26).

Same experimental conditions. Observation time $\Delta T = 6$ s (interval *ab);* spatial interval occupied by the vortex: ~ 9 m. Averaging time: 0.025 s, spatial averaging interval: 0.04 m. At wavelength $\lambda = 17$ m, the linear size of the vortex should be ~

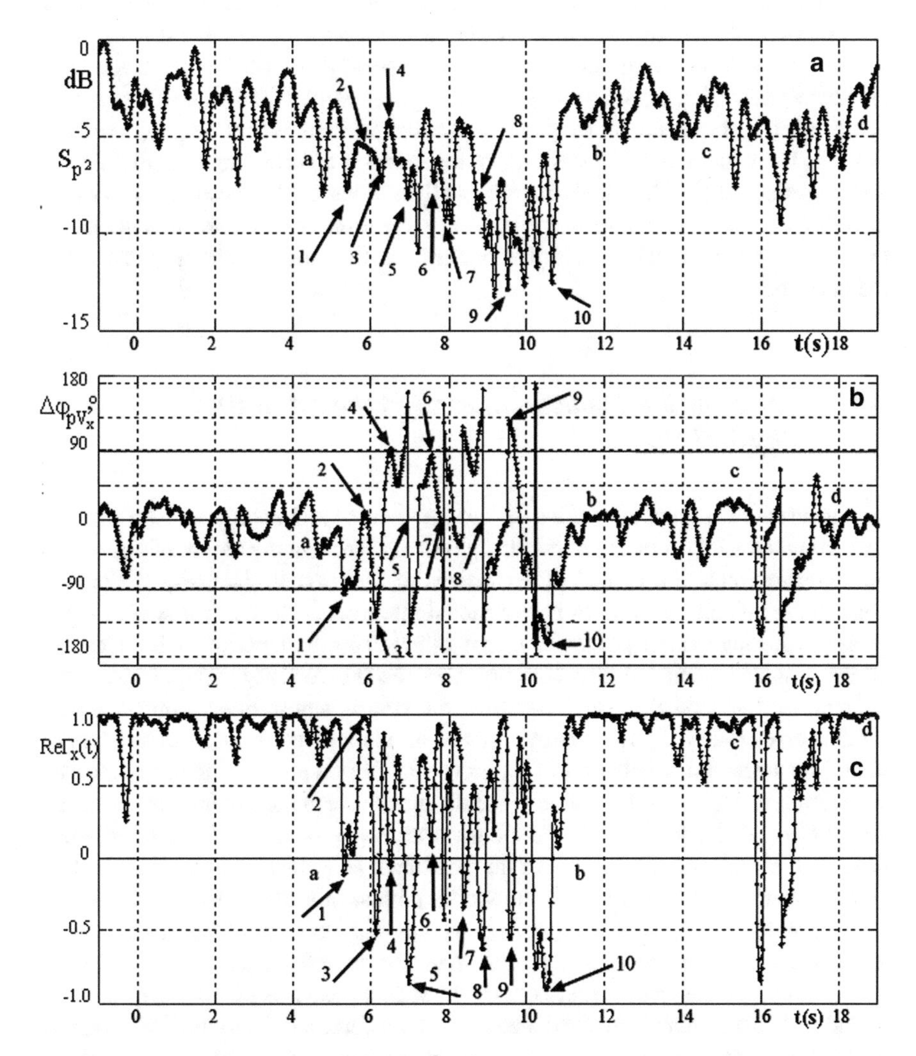

Fig. 4.26 Vortex δ. Time dependence: **a** $p^2(t)$; **b** $\Delta\varphi_{pV_x}(t)$, **c** Re $\gamma_x(t)$. Frequency: 88 Hz. Averaging time: 0.025 s

1.7 m. The 1.7 m interval is equivalent to ~ 42 spatial averaging intervals, ensuring high-accuracy scanning of the vortex and its movement.

It can be presumed that ~ 5.3 vortices are observed inside a 9 m interval. It is known that neighbouring vortices should normally be opposite in sign, ' + ' to '–'. We will begin out analysis with point 1 (Fig. 4.26): $p(t)$ isn't small; $\Delta\varphi_{pV_x}(t) = \pi/2$; Re $\Gamma_x(t) = 0$. As a result, point 1 is near a saddle according to the vortex criteria. Next, point 2: $p(t)$ grows to the level of constructive interference; $\Delta\varphi_{pV_x}(t) = 0°$; Re $\Gamma_x(t) = + 1.0$. The receiver has left the vortex region and is now in the area of constructive interference.

Further in time, $p^2(t)$ fluctuates as it moves from the region of destructive to constructive interference; $\Delta\varphi_{pV_x}(t)$ varies within 2π; Re $\Gamma_x(t)$ fluctuates from + 1.0 to – 1.0 and passes through zero. The change of phase difference $\Delta\varphi_{pV_x}(t)$ is clockwise; therefore, we are dealing with a single '–' vortex. The total phase advance is 8π over 6 s; displacement frequency is 1.5 Hz. We will denote $\Delta\varphi_{pV_x}(t) = 2\pi \cdot n$, where n is the number of jumps on passing through a vortex horizontally. All the above vortices can be classified according to their n. Vortex β has $n = 1$; vortex α, $n = 2$; vortex γ, $n = 3$; and vortex δ, $n = 4$. In a ' + ' vortex, energy flows anticlockwise; and in a '–' vortex, clockwise. In the former case, the saddle is above the centre; in the latter, below.

4.4.2 *Vortex of the Acoustic Intensity Vector as a Real Physical Object*

Vector-phase experimental studies of low-frequency tonal energy transfer in a real shallow water waveguide have revealed physical objects in the region of destructive interference which are identified as vortices of the acoustic intensity vector. The mechanism behind the vortices is intermodal interference. The movement of signal energy in the outer vicinity of a vortex, as it follows from the experiment, is large-scale vortex in nature, giving rise to vorticity. Vortices and vorticity make up the vortex structure of the acoustic field. Therefore, the vortex, whose linear dimensions are ~ 0.2λ, gives rise to vortex energy transfer in a significant portion of the waveguide. The structural foundation of an individual vortex are its singular points: the dislocation (centre) and the saddles. The dislocations and the saddles are connected with each other and form a stable topological structure—a vortex. In the interference process, the centres are 'coupled' to the minima of acoustic pressure, while particle velocity minima are 'coupled' to the saddle points. In a real interference process, displacement of a pressure minimum, as well as displacement of a particle velocity minimum, is a natural process caused by hydrodynamic fluctuations in the medium. Displacement of the pressure zero causes the saddle points, as [8, 10] suggest, to 'do a kind of "dance" about the field's zeroes'. Therefore, by observing the dynamics of a vortex over time, one can follow the fluctuations of the interference field and trace them back to their causes. The dynamic big picture of vortex energy transfer in the real waveguide in the far field of a source is as follows. The regular arrangement of vortices along the waveguide axis creates a regular alternating structure of energy flows in the vertical plane. Energy streamlines bending around vortices create an eddy field with a nonzero rotation. Therefore, vortices and vorticity create a common vortex structure of acoustic signal energy transfer in a shallow water waveguide.

4.5 Conclusions

Modal structure and vortices of the acoustic intensity vector determine the structure of acoustic field and the transfer of energy in the shallow water waveguide. Because vortices result from modal interference, horizontal oscillatory displacements of vortices can be viewed as a result of reconfiguration of the modal structure. It has been demonstrated theoretically that vortices can travel considerable distances (~ 150 m) in rising tides of 0.3 m [10].

In our experiments, we observed only oscillatory displacements of vortices over distances up to signal wavelength under constant mean sea level. The first and the second experiment show that vortex transfer of signal energy in the shallow water waveguide occurs at all depths in the waveguide, regardless of whether vortices exist at a given depth or not. Hence, it follows that the movement of energy in the shallow water waveguide has a vortex nature throughout the waveguide water column.

It has been experimentally shown that the mode theory model of low-frequency signal energy propagation in a shallow water waveguide has its limitations, and it is necessary to take into account eddy transport of energy and its relationship with hydrodynamic processes. A stable physical object, the vortex can be employed as a research tool. Building a complete picture of vortex energy transfer throughout the depth of the waveguide requires experimental studies with vertically spaced combined receivers.

References [14–16, 24–30] cite additional sources relevant to this problem.

References

1. M.A. Isakovich, *General Acoustics* (Nauka, Moscow, 1973). (in Russian)
2. J. Mann, T. Tichy, A.J. Romano, Instantaneous and time-averaged energy transfer in acoustics fields. J. Acoust. Soc. Am. **82**(4), 17–30 (1987)
3. S.M. Rytov, *Introduction to Statistical Radiophysics* (Nauka, Moscow, 1976). (in Russian)
4. V.A. Shchurov, V.P. Kuleshov, E.S. Tkachenko, Phase spectra of interference of a broadband surface source in shallow water, in *Proceedings of the 22nd Session of Russian Acoustic Society and session of the Acoustics Scientific Council of the Russian Academy of Sciences*, vol. 2 (GEOS, Moscow, 2010), pp. 248–251. (in Russian)
5. V.A. Shchurov, V.P. Kuleshov, A.V. Cherkasov, Vortex properties of the acoustic intensity vector in shallow water. Akusticheskij Zhurnal (Acoust. Phys.) **57**(6), 837–843 (2011). (in Russian)

6. V.A. Shchurov, A.S. Lyashkov, On some features of energy characteristics of shallow-water acoustic interference field. Akusticheskij Zhurnal (Acoust. Phys.) **59**(4), 459–468 (2013). (in Russian)
7. V.A. Zhuravlev, I.K. Kobozev, Yu.A. Kravtsov, Energy flows in the vicinity of dislocations of the wavefront phase field. JETP **104**(5(11)), 3769–3783 (1993). (in Russian)
8. J.F. Nye, M.V. Berry, Dislocations in wave trains. Proc. R. Soc. Lond. **336**(1605), 165–190 (1974)
9. N.B. Baranova, B.Y. Zeldovich, Dislocations of the wavefront surface and amplitude zeros. JETP **80**(5), 1789–1797 (1981). (in Russian)
10. V.A. Zhuravlev, I.K. Kobozev, Yu.A. Kravtsov, Dislocation of the phase front in the ocean waveguide and their manifestation in acoustic measurements. Akusticheskij Zhurnal (Acoust. Phys.) **35**(2), 260–265 (1989). (in Russian)
11. L.M. Brekhovskikh, Yu.P. Lysanov, *Theoretical Foundations of Ocean Acoustics* (Gidrometeoizdat, Leningrad, 1982). (in Russian)
12. V.A. Eliseevnin, Yu.I. Tuzhilkin, Acoustic power flux in the waveguide. Akusticheskij Zhurnal (Acoust. Phys.) **47**(6), 781–788 (2001). (in Russian)
13. A.N. Zhukov, A.N. Ivannikov, V.I. Pavlov, On identification of multipole sound sources. Akusticheskij Zhurnal (Acoust. Phys.) **36**(3), 447–453 (1990). (in Russian)
14. C.F. Chien, Singular points of intensity stream lines in two-dimensional sound field. J. Acoust. Sos. Am. **101**(2), 705–712 (1997)
15. O.R. Lastovenko, V.A. Lisyutin, A.A. Yaroshenko, Features of vector acoustic fields in shallow water waveguides, in *Consonance-2011 Acoustic Symposium*, pp. 188–193
16. D.R. DallOsto, P. Dahl, Properties of acoustic intensity vector field in a shallow water waveguide. J. Acoust. Sos. Am. **131**(3), 2023–2035 (2012)
17. V.A. Shchurov, Peculiarities of real shallow sea wave-guide vortex structure. J. Acoust. Soc. Am. **145**(1), 525–530 (2019)
18. V.A. Shchurov, The dynamics of low-frequency signal acoustic intensity vector vortex structure in shallow sea. Chinese J. Acoust. **38**(2), 113–131 (2019)
19. I. Tolstoy, C.S. Clay, *Ocean Acoustics*, Russian. (Mir, Moscow, 1969)
20. A.N. Samchenko, I.O. Yaroshchuk, Acoustic parameters of unconsolidated bottom sediments of the Peter the Great Bay (Sea of Japan). Vestnik Far Eastern Branch Russian Acad. Sci. **5**, 130–136 (2017). (in Russian)
21. V.A. Shchurov, A.S. Lyashkov, S.G. Shcheglov, E.S. Tkachenko, G.F. Ivanova, A.V. Cherkasov, Local structure of the shallow sea interference field. Underwater Invest. Robot. **1**(17), 58–67 (2014). (in Russian)
22. V.A. Shchurov, Large- and small-scale acoustic vortices intensities assessment, in *The 5th Pacific Rim Underwater Acoustics Conference. 2015. Proceedings of Meetings on Acoustics, layered Ocean* (2016), pp. 1–7
23. V.G. Petnikov, V.A. Popov, AYu. Shmelev, Dislocation tomography of the ocean: a new method of acoustic diagnostics (in Russian). Akusticheskij Zhurnal (Acoust. Phys.) **39**(4), 764–765 (1993)
24. V.P. Dzyuba. *Scalar-Vector Methods of Theoretical Acoustics* (Dalnauka, Vladivostok, 2006). (in Russian)
25. D.R. DallOsto, Properties of the Acoustic Vector Field in Underwater Waveguides. A dissertation for the degree of Doctor of Philosophy, University of Washington, 2013
26. V.V. Borodin, V.A. Zhuravlev, I.K. Kobozev, Y.A. Kravtsov, Averaged characteristics of acoustic fields in ocean waveguidesx. Akusticheskij Zhurnal (Acoust. Phys.) **38**(4), 601–608. (in Russian)
27. V. Shchurov, Comparison of the vorticity of acoustic intensity vector at 23 Hz and 110 Hz frequencies in the shallow sea. Appl. Phys. Res. **3**(2), 179–189 (2011)
28. B.A. Kasatkin, N.V. Zlobina, S.B. Kasatkin, *Model Problems in Acoustics of Layered Media* (Dalnauka, Vladivostok, 2012). (in Russian)
29. C.S. Clay, H. Medwin, *Acoustic Oceanography*, Russian. (Mir, Moscow, 1980)
30. V.N. Kulakov, N.E. Maltsev, S.D. Chuprov, On excitation of a group of modes in a layered ocean. Akusticheskij Zhurnal (Acoust. Phys.) **29**(1), 74–79 (1983)

Chapter 5
Observing Weak Signal in Diffuse, Partially Coherent and Coherent Acoustic Noise

5.1 Introduction

The main objective of this chapter is to present new information criteria that expose a moving underwater source based on experimental investigations into the space-time structure of the vector acoustic fields of signal and noise. Vector properties of the underwater acoustic field described in Chaps. 1–4 can be used to build a system of algorithms to solve detection and classification problems at a whole new technological level.

In the USSR (Russia), noise immunity potential of the combined receiver provoked lengthy debate in the acoustic research community. Experiments in shallow water and in the deep open ocean proved that combined acoustic receiving systems are effective. A field experiment in the deep open ocean proved that noise immunity of an individual combined receiver relative to the square-law detector of a single hydrophone can reach ~ 15–20 dB in a diffuse (isotropic) field and ~ 30 dB in an anisotropic field [2]. This finding sparked an argument in the *Akusticheskij Zhurnal* (*Acoustical Physics*) [3, 4]. The opponents, asserting their theoretical and model-based views while ignoring the author's experimental results [2], argued that multiplicative data processing is unable to achieve a gain in noise immunity relative to the square-law detector [3]. The author believes that the error of the authors of [3, 5–7] was that they viewed one channel of the vector receiver as an individual hydrophone, albeit with a cosine directivity pattern. Naturally, with such multiplicative processing, noise immunity of the receiving system in a diffuse field was no more than 3 dB. Thence the opponents concluded that multiplicative processing is inferior to additive. This opposition to the vector-phase method significantly held back its adoption in applied hydroacoustics. Multiplicative processing of the four components of the field $p(t)$, $\mathrm{V}(t)\{V_x(t), V_y(t), V_z(t)\}$ is a cross-correlation analysis that calculates the Umov vector, the intensity vector and the differential phase relations while suppressing the noise of the field's diffuse component.

An individual four-component combined receiver, small in wavelength terms, has a normalised spherical directivity pattern with a spatial coverage over the 4π solid

V. A. Shchurov, *Movement of Acoustic Energy in the Ocean*,
https://doi.org/10.1007/978-981-19-1300-6_5

angle independent of angles θ and φ,

$$R^2(\theta, \varphi) = R_x^2(\theta, \varphi) + R_y^2(\theta, \varphi) + R_z^2(\theta, \varphi) \tag{5.1}$$

where $R_x = R_0 \sin\theta \cos\varphi$, $R_y = R_0 \sin\theta \sin\varphi$, $R_z = R_0 \cos\theta$; R_0 is axial sensitivity of channels *x, y, z*; φ is azimuthal angle measured from the *x* axis; and θ is the polar angle. Individual combined receivers can be combined into detection arrays [13–15]; they can also be conveniently carried by gliders [16].

Subsequent theoretical and experimental work [8–11] fully confirmed the findings of [2], which in essence is the first theoretical and experimental work to determine noise immunity of an individual combined receiver in complex acoustic fields.

In the USA, France and Japan, the vector-phase approach is used in sonar detection systems. The US Navy has been using an acoustic vector-phase buoy called DIFAR (Directional Low-Frequency Analysis and Recording) to detect submarines since the 1950s. Recent years have seen much research on introducing combined receivers into towed and stationary antennas, and especially on placing vector assets on unmanned marine platforms such as gliders. Appendix II lists some patents and papers on the subject published in international scientific press in recent years; of these, 20 are patents (17 USA, Navy; 3 China) and 74 are papers published in JASA, IEEE and presented at international conferences. The patents and papers look at the principles of constructing vector receivers, vector antennas and processing algorithms. Nearly all the work is sponsored by the US Navy, indicating that vector systems are being used for the purposes of the US Navy.

This chapter presents the theory of noise immunity of an individual combined receiver and estimates its gain in noise immunity over the square-law detector. Using real spectra for illustration, we present the characteristics of underwater ambient noise and signal in the horizontal and vertical planes. The compensation phenomenon (Chap. 4) and the method of directivity pattern rotation form the basis for the vector-phase sonar. We present signal processing flowcharts that use the FFT and the Hilbert transform.

5.2 Noise Immunity of an Individual Combined Receiver in the Case of a Tonal Signal

Considered as a vector field, the acoustic field takes on a number of crucial properties that are absent from the scalar description. For example, one such property is compensation of reciprocal energy fluxes, which can be used to build target detection and classification algorithms.

A crucial parameter of the vector acoustic field are phase differences between acoustic pressure and components of the particle velocity ($\Delta\varphi_{pV_x} = \varphi_p - \varphi_x$, $\Delta\varphi_{pV_y} = \varphi_p - \varphi_y$, $\Delta\varphi_{pV_z} = \varphi_p - \varphi_z$) and between components of particle velocity ($\Delta\varphi_{xy} = \varphi_x - \varphi_y$, $\Delta\varphi_{xz} = \varphi_x - \varphi_z$, $\Delta\varphi_{yz} = \varphi_z - \varphi_{yz}$). It is an experimental fact that differential phase relations are the most consistent and robust characteristics of acoustic field in statistical signal processing.

Real underwater dynamic ambient noise contains a diffuse and a coherent component. The coherent component of dynamic noise can be easily identified and is usually broadband. Nearby shipping noise is partially or fully coherent. These studies should focus on investigating the features of acoustic energy transfer in the shallow water waveguide from surface and underwater sources.

Statistical signal processing is organised as follows. The acoustic field is considered stationary and ergodic and the signal monochromatic. Fast Fourier transform is used in the frequency domain, and the Hilbert transform in the time domain. We use the concept of coherence and partial coherence of reciprocal acoustic variables. The degree of coherence of useful signal or noise is determined in the frequency domain by the function $\gamma^2_{pV_i}(f)$ and in the time domain by the function $\operatorname{Re}\Gamma_{pV_i}(t)$, where $i = x, y, z$. When $\gamma^2_{pV_i}(f, t_0) = +1$, $\operatorname{Re}\Gamma_{pV_i}(f_0, t) = \pm 1$ and $\operatorname{Im}\Gamma_{pV_i}(f_0, t) = 0$, we consider the signal to be fully coherent. When $\gamma^2_{pV_i}(f, t_0) < +1$, $-1 < \operatorname{Re}\Gamma_{pV_i}(f_0, t) < +1$ and $-1 < \operatorname{Im}\Gamma_{pV_i}(f_0, t) < +1$, we consider the signal or the noise to be partially coherent. If $\gamma^2_{pV_i}(f, t_0) = 0$, $\operatorname{Re}\Gamma_{pV_i}(f_0, t) = 0$and $\operatorname{Im}\Gamma_{pV_i}(f_0, t) = \pm 1.0$, the acoustic field is diffuse.

Papers [1, 2, 8] examine the process that determines the signal-to-noise ratio SNR(PV) of the combined four-component acoustic receiver during multiplicative signal processing. Mathematical processing of the four acoustic channels $p(t)$, $V_x(t)$, $V_x(t)$, $V_z(t)$ located at one point in space with a common phase centre consists in auto- and cross-correlation of the four components with a time shift $t = 0$. The result shows how mutually coherent the real characteristics of the acoustic field $p(t)$, $V_x(t)$, $V_x(t)$, $V_z(t)$ are and how signal levels compare with underwater ambient noise and coherent or partially coherent interference noise. Papers [2, 8] show that upon averaging, the components of isotropic ambient noise $\langle p(t)V_x(t)\rangle_t$, $\langle p(t)V_y(t)\rangle_t$, $\langle p(t)V_z(t)\rangle_t$ can 'drop' by about 16–20 dB relative to $\langle p^2(t)\rangle_t$. The averaging decay of energy flux density of isotropic ambient noise is approximated by $b/\sqrt{\Delta f T}$, where b is a coefficient; Δf is the frequency band; and T is the averaging time [1, 2]. Variance $\sigma^2(pV)$ is expressed through pressure variance σ^2_p and particle velocity variance σ^2_V at the output of the measuring system as follows [8]:

$$\sigma^2(pV) = \frac{1}{4}\left(\sigma^2_{p,N} + \sigma^2_{V,N} + 2\sigma^2_{p,N}\sigma^2_{V,N}\right), \tag{5.2}$$

where γ is the correlation coefficient between p and V. In the case of dynamic surface noise

$$\gamma = 0 \text{ and } \sigma^2(pV) = \frac{1}{4}\left(\sigma^2_{p,N} + \sigma^2_{V,N}\right). \tag{5.3}$$

Since the pressure p channel has a circular directivity pattern and the particle velocity V channel a cosine one, the ratio of variances σ^2_p and σ^2_V in an isotropic field is determined by the directivity factor of the V channel that equals 3. In multiplicative processing, excess $\langle (pV_i)_S \rangle_t$ signal is uniquely determined against the background of reduced noise level $\langle (pV_i)_N \rangle$, the isotropic field and the fluctuation component σ^2_{pV}, where $i = x, y, z$ at the output channel of the combined receiver. Signal excess in the pressure channel is estimated as $\langle P_S^2 \rangle - \langle P_N^2 \rangle$ relative to fluctuations $\tilde{p}_N^2$. For a component of the energy flux density vector, excess is estimated as $\langle (pV_i)_S \rangle - \langle (pV_i)_N \rangle$ against the background of fluctuations $\widetilde{pV}_{i,N}$. In the case of a deterministic source, $\langle P_S^2 \rangle = \langle (pV) \rangle_S$. Therefore, noise energy flux density $\langle (pV_i)_N \rangle$ and fluctuations of the noise energy flux $\widetilde{pV}_{i,N}$ should be compared with the corresponding value at the outlet of the pressure receiver

$$\Delta = p_N^2 + \tilde{p}_N^2/(pV_i)_N + (\widetilde{pV}_i)_N \tag{5.4}$$

As shown in [8], signal-to-noise ratio for pressure, particle velocity and energy flux density vector can be written as:

$$\begin{aligned} \mathrm{SNR}(p^2)_N &= k_p^2/\sqrt{1 + (1 + 2k_p^2)/b\tau}; \\ \mathrm{SNR}(V^2)_N &= \beta^2 k_p^2 \sqrt{1 + (1 + \beta^2 k_p^2)/b\tau}; \\ \mathrm{SNR}(pV)_N &= \beta^2 k_p^2 \sqrt{b\tau}/\sqrt{1 + k_p^2(1 - \beta)} \end{aligned} \tag{5.5}$$

where $k_p^2 = p_S^2/p_N^2, k_V^2 = V_S^2/V_N^2, (\beta k_p)^2 \approx (p_S^2/p_N^2)(p_N^2/V_N^2), b = 2\pi \Delta f$ and Δf is the analysis bandwidth. With increasing averaging time t, these quantities behave as follows:

$$\mathrm{SNR}(p^2) \to k_p^2; \ \mathrm{SNR}(V^2) \to k_V^2; \ \mathrm{SNR}(pV); \text{ goes as } \sqrt{b\tau}.$$

We should note that the Cron–Sherman isotropic field model used in [8] doesn't accurately describe the real vector field of dynamic acoustic noise. The model assumes that the dynamic noise of the sea surface can be modelled by a circle with independent sound sources evenly distributed throughout. An isotropic noise field will exist in the centre of the circle. However, as shown in [1], sea surface waves are a source of coherent noise whose direction of propagation is determined by the direction of surface waves and the direction of surface wind. Neither is the volumetric noise model, presented in [8] as a sphere uniformly filled with independent sound sources, accurate, since there are vertical coherent flows of noise energy.

The diffuse field concept should be used for the real noise field, since the degree of isotropy or degree of coherence of the noise field varies with azimuthal angle φ and polar angle θ, meaning that the ratio Δ (5.4) can vary with φ and θ. Papers [1, 2] use vector-phase processing of experimental data to demonstrate that signal-to-noise ratio SNR$(pV)_N$ in a diffuse field achieves a gain on the order of 15–16 dB relative to an individual hydrophone SNR$(p^2)_N$. Naturally, the resulting gain is determined by the 'jump' of Δ (5.4) of the acoustic field—that is to say, by how much of the total field is diffuse field.

In an anisotropic field of local interference noise, the gain can be made as high as 26–28 dB (equal to the sensitivity relation coefficient of an individual channel of the vector receiver) by pointing the sensitivity minimum of a channel of the vector receiver towards the source of interference noise according to the formula [1]:

$$u_x^{'} = u_x \cos\varphi_0 + u_y \sin_0,\ u_y^{'} = -u_x \sin\varphi_0 + u_y \cos_0,$$

where u_x and u_y are electrical signals from channels x and y recorded on magnetic tape in the experiment; $u_x^{'}$ and $u_y^{'}$ are electrical signals from channels x and y rotated through the angle φ_0 about the radiation axis.

Figure 5.1 presents the spectra of magnitude of acoustic pressure $S_{p^2}(f)$ and components of coherent power $S_{pV_x}(f) = \gamma^2_{pV_x}(f)S_{p^2}(f)$, $S_{pV_y}(f) = \gamma^2_{pV_y}(f)S_{p^2}(f)$, $S_{pV_z}(f) = \gamma^2_{pV_z}(f)S_{p^2}(f)$. The averaging time is 128 s. Analysis

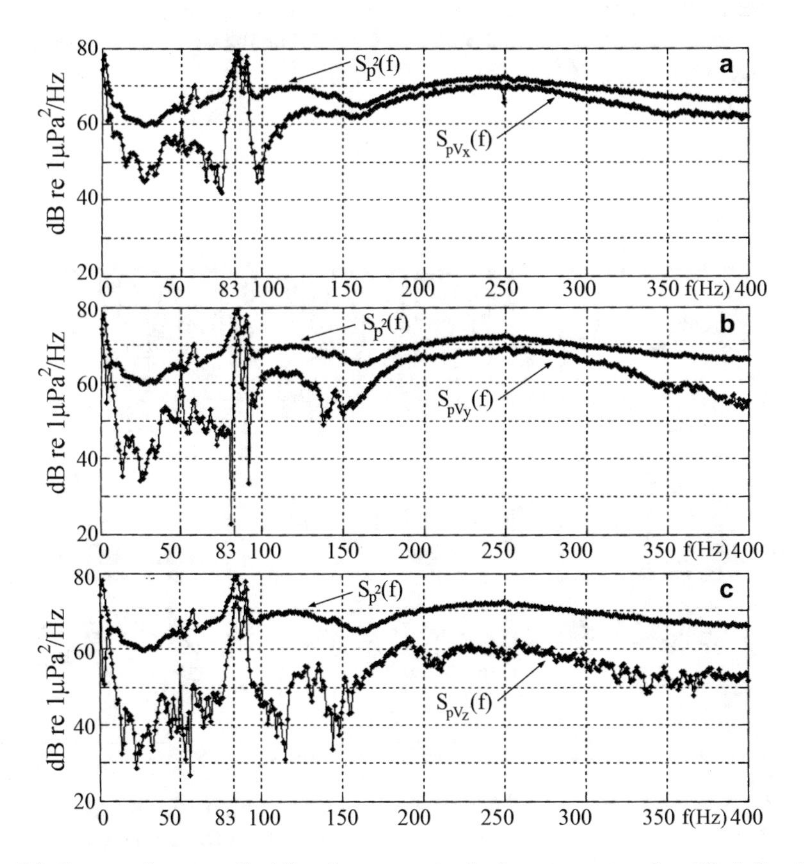

Fig. 5.1 Spectra of pressure $S_{p^2}(f)$ and components of coherent power: **a** $S_{pV_x}(f)$, **b** $S_{pV_y}(f)$, **c** $S_{pV_z}(f)$. The tone frequency is 83 Hz. Analysis bandwidth is 1 Hz. Averaging time: 128 s. The underwater source is ~ 8 km away

bandwidth is 1 Hz. The significant 128 s averaging time was chosen so as to suppress the diffuse (incoherent) noise and the partially coherent interference noise from random sources as much as possible [1]. Measurements were made in the Peter the Great Gulf, Sea of Japan, in 60 m of water. Sea state was 3 on the Beaufort scale. Figure 5.1 shows that when coherent power is measured as a function of frequency, the noise level 'drops' by 3–20 dB relative to the pressure noise in all channels of the combined receiver. Coherent power excess relative to the *p* channel of the 83 *Hz* signal over the noise and interference for channels *x, y, z* is at least 20dB; *p* channel excess is 10dB. In the spectrum of the underwater ambient noise, dynamic noise gently peaks around 150–350 Hz due to surface waves [1]. The coherent power of *x*, *y* and *z* components is 3–10 dB below the pressure noise across all frequencies due to suppression of the diffuse component of ambient noise.

Figure 5.2 shows spectra of vector characteristics of acoustic field $S_{p^2}(f), S_{pV_i}(f)$, $\gamma^2_{pV_i}$, $\Delta\varphi_{pV_i}$; the averaging time is 20 s. The combined receiver (15 m deep) is ~ 15 km away from the source (~20 m deep). The harmonic radiation frequency was 83 Hz. The decibel scale is arbitrary. Pressure signal-to-noise excess doesn't exceed 5 dB. From Fig. 5.2 it follows that $\gamma^2_{pV_x}(f) = 0.45$; $\gamma^2_{pV_y}(f) = 0.6$; $\gamma^2_{pV_z}(f) = 0.1$ 0.1—that is, the 83 Hz signal appears partially coherent at the point of reception, while dynamic noise is significantly more coherent than the signal above 100 Hz, its coherence reaching $\gamma^2_{pV_x}(f) = 0.8$.

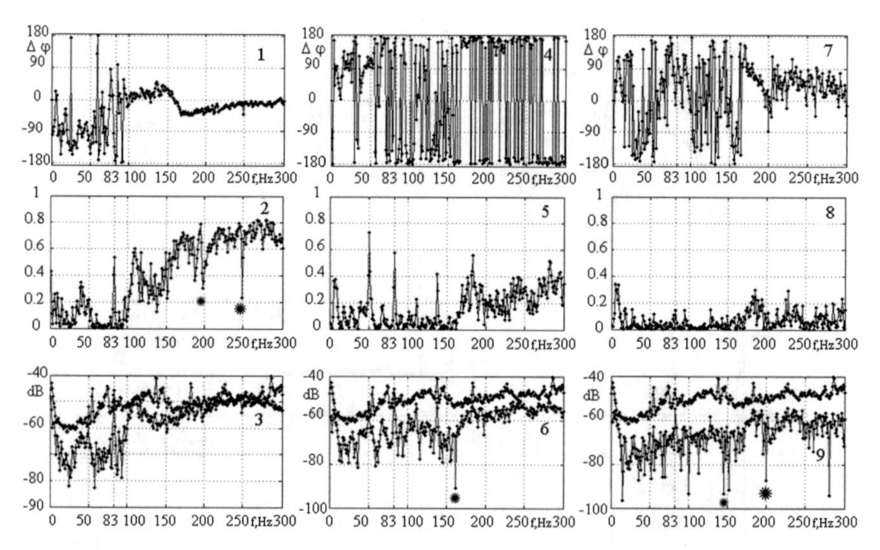

Fig. 5.2 Spectra: 1 $\Delta\varphi_{pV_x}$, 2 $\gamma^2_{pV_x}$, 3 S_{p^2} and S_{pV_x}; 4 $\Delta\varphi_{pV_y}$, 5—$\gamma^2_{pV_y}$, 6 S_{p^2} and S_{pV_y}; 7 $\Delta\varphi_{pV_z}$, 8 $\gamma^2_{pV_z}$, 9 S_{p^2} and S_{pV_z}. Averaging time: 20 s. Analysis bandwidth: 1 Hz. Local depth: 30 m. The 83 Hz underwater source was ~ 15 km away

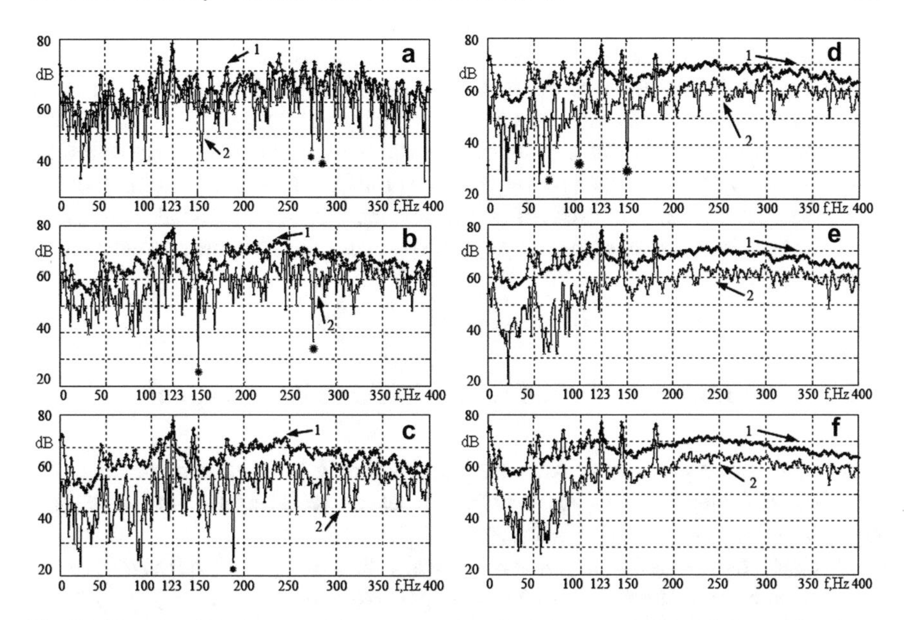

Fig. 5.3 Spectra of $S_{p^2}(f)$ and of the y component of coherent power $S_{pV_y}(f)$. Averaging times: **a** 1 s; **b** 2 s; **c** 4 s; **d** 8 s; **e** 16 s; **f** 32 s. Key: **a** $S_{p^2}(f)$, 2—$S_{pV_x}(f)$. Analysis band: 1 s. Asterisk (*) marks spectral lines that result from compensation of reciprocal energy fluxes. Decibel levels are relative to 1 μPa^2/Hz. The 123 Hz underwater source is ~ 12 km away

However, the signal/noise excess of coherent power at 83 Hz relative to pressure is, by channel: x 20dB, y 15dB, z 10dB. Phase differences are $\Delta\varphi_{pV_x} = 180°$, $\Delta\varphi_{pV_y} = 180°$, $\Delta\varphi_{pV_z} = 100°$. This combination of characteristics indicates that the 83 Hz signal is observed. $S_{p^2}(f)$ exceeds the noise by no more than 5 dB. Note the discrete lines shaped like 'dips', some of them marked with an asterisk * on the spectra $\gamma^2_{pV_x}(f)$, $S_{pV_y}(f)$ and $S_{pV_z}(f)$. The origin of the discrete 'dips' lies in compensation of reciprocal energy fluxes [13].

Figure 5.3 presents spectra $S_{p^2}(f)$, $S_{pV_i}(f)$ $(i = x, y, z)$ at various averaging times. Underwater source frequency is 123 Hz. Figure 5.2 shows that depending on the frequency and averaging time, coherent power recedes 10–30 dB relative to the spectrum $S_{p^2}(f)$. The greatest decline of the coherent power spectrum is observed at frequencies up to 150 Hz (up to 30 dB). At higher frequencies, dynamic noise gently peaks (around 150–350 Hz) due to surface waves.

Spectral levels of $S_{p^2}(f)$ in Fig. 5.3d–f are almost identical, meaning that averaging over 8 s is sufficient for a statistically robust spectrum $S_{p^2}(f)$; by contrast, a statistically robust spectrum $S_{pV_y}(f)$ requires an averaging time of 16 s, since the $S_{pV_y}(f)$ spectra in Fig. 5.3e, f are identical. This mechanism of noise immunity of the combined receiver has been discussed in scientific literature for over 30 years. Real experimental data suggest that depending on the type of mathematical processing, noise immunity of the combined receiver can range from 5 dB (in an isotropic field) to ~ 25–30 dB (in an anisotropic field). However, due to its vector nature, in a complex acoustic field, with flux density vectors from various sources adding together and a weak signal (at the level of the p channel noise), the desired signal, that is, its coherent component, may be compensated by reciprocal fluxes of noise or interference energy. In this case (of a complex acoustic field), the methodology discussed above is not applicable. It then follows that the problem of detecting a weak signal in a complex acoustic field consists of two stages: the first one is a signal with at least a + (2–3) dB excess over p; and the second is when the signal power is below the p level.

5.3 Noise Immunity in the Case of a Broadband Signal

Figure 5.4 shows the path of a diving boat used as a broadband projector in the 50–1000 Hz frequency band. The boat moved past the receiving system at a constant speed of ~ 9 kts on a constant heading.

Passage characteristics are shown in Figs. 5.5, 5.6, 5.7, 5.8, 5.9 and 5.10. At different averaging times 5, 10 and 15 s, passage characteristics $E_{\mathrm{p}} = pp^*$ are compared against coherent power $pV_i^*(i = x, y, z)$ in the frequency band $\Delta f = 353 - 397\text{Hz}(\Delta f = 44\text{Hz}))$. Numbers 1–8 in Figs. 5.5, 5.7 and 5.9 have these meanings: 1 is underwater ambient noise level for E_p in the absence of the diving boat; 2 is noise for $pV_i^*(i = x, y, z)$. The number 3 marks the section of the passage characteristic E_p where there are no large enough changes in E_p but where pV_i^*

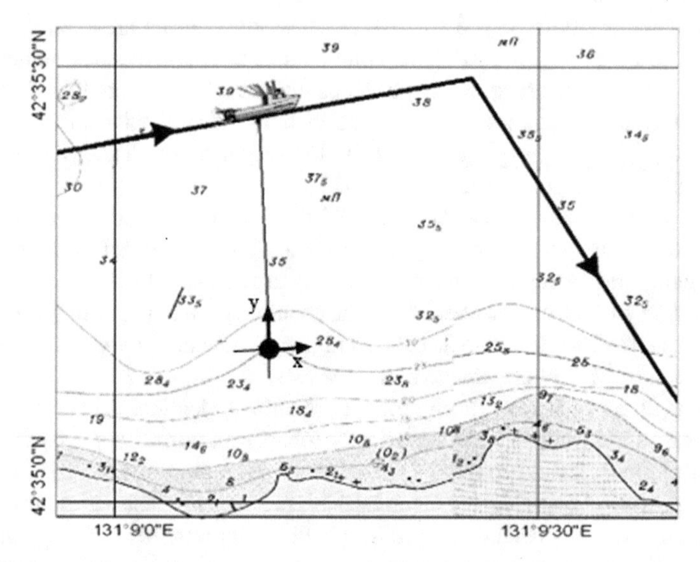

Fig. 5.4 Path travelled by the diving boat relative to the receiving system. The chart shows a part of the path with the vessel moving from ~ 10 km away towards the receiving system. Distance abeam is ~ 300 m

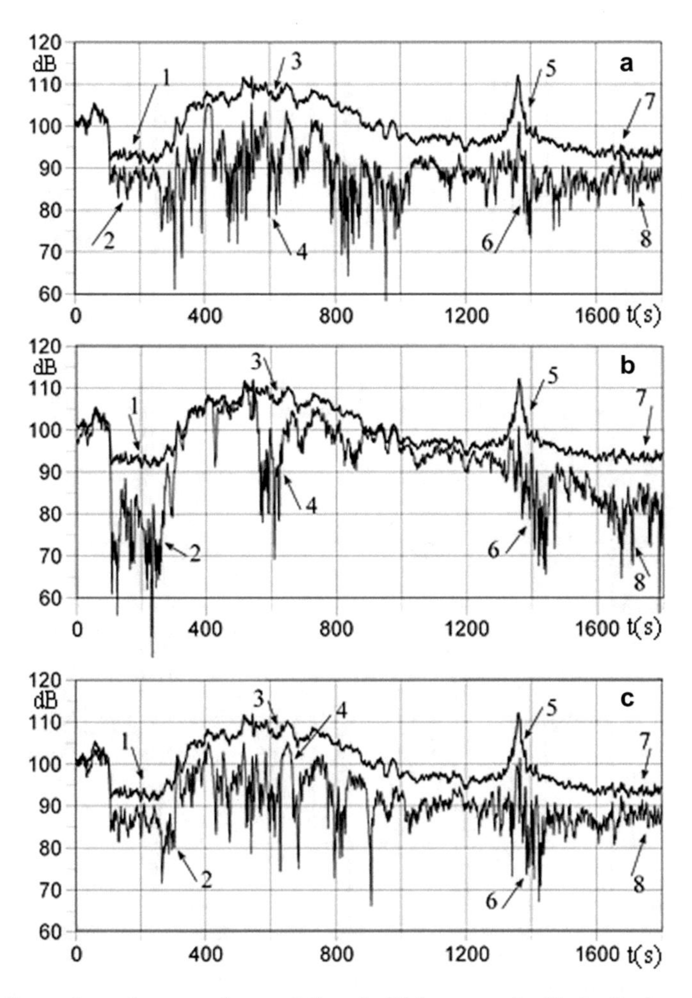

Fig. 5.5 Comparison of passage characteristics of $pV_i^*(i = x, y, z)$ with the E_p characteristic: **a** pp^* (top) and $pV_x{}^*$; **b** pp^* and pV_y^*; **c** pp^* and pV_z^*. Vessel abeam at $t = 600$ s. $R_{\text{abeam}} \approx 300$ m. Maximum distance of ~ 7 km at $t = 1800$ s. Averaging time is 5 s, $\Delta f = 353 - 397$ Hz

(number 4) experiences two phenomena at once: interference and compensation of energy fluxes with a significant change of level (up to 20dB). The number 5 marks a change of E_p when passing near the receiving system; the number 6 marks changes in pV_i^* associated with compensation of energy fluxes during the passage. The numbers 7 and 8 mark the level of E_p and pV_i^*, respectively, when the diving boat is ~ 7 km away. Figures 5.6, 5.8 and 5.10, numbers 1 and 2, 3 and 4, 5 and 6, 7 and 8 denote phase differences $\Delta\varphi_{pV_i} = \Delta\varphi_p - \Delta\varphi_z$ $(i = x, y, z)$ for the passage characteristics in Figs. 5.5, 5.7 and 5.9. Figures 5.5, 5.6, 5.7, 5.8, 5.9 and 5.10 are shown at averaging times increasing from 5 to 15 s to analyse the variability of vector field structure compared with the scalar field of acoustic pressure. The

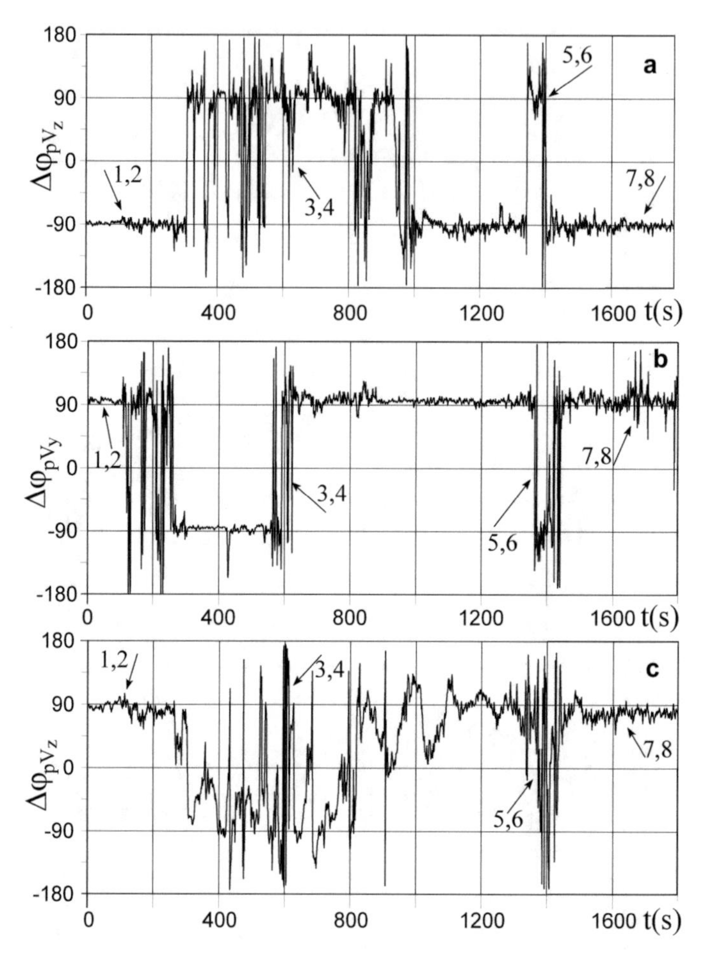

Fig. 5.6 Differential phase passage characteristics: **a** $\Delta\varphi_{pV_x}$; **b** $\Delta\varphi_{pV_y}$; **c** $\Delta\varphi_{pV_z}$ for Fig. 5.5. Same conditions as in Fig. 5.5

analysis was done in the frequency band $\Delta f = 353 - 397$ Hz. The orthogonal *x*, *y* and *z* components are components of particle acceleration.

Figures 5.5, 5.7 and 5.9 show that while fluctuations of E_{p} never exceed 6 dB (see e.g. Fig. 5.9), the pV_{i}^* curves are jagged and fluctuate as much as 40 dB. Noteworthy is the following: the number 1 marks the situation where the diving boat is adrift, and it can be assumed that in this interval the only thing recorded is background noise in the acoustic pressure channel; number 2 marks the noise level in the pV_{i}^* channels; these are significantly (at least 10–20 dB) lower than pp^*. Difference of levels 1 and 2 among orthogonal channels *x, y, z* indicates that the acoustic field is significantly anisotropic in three dimensions.

Figures 5.6, 5.8 and 5.10 show differential phase characteristics $\Delta\varphi_{pV_x} = \varphi_p - \varphi_x$, $\Delta\varphi_{pV_y} = \varphi_p - \varphi_y$, $\Delta\varphi_{pV_z} = \varphi_p - \varphi_z$. Because the phase difference between the

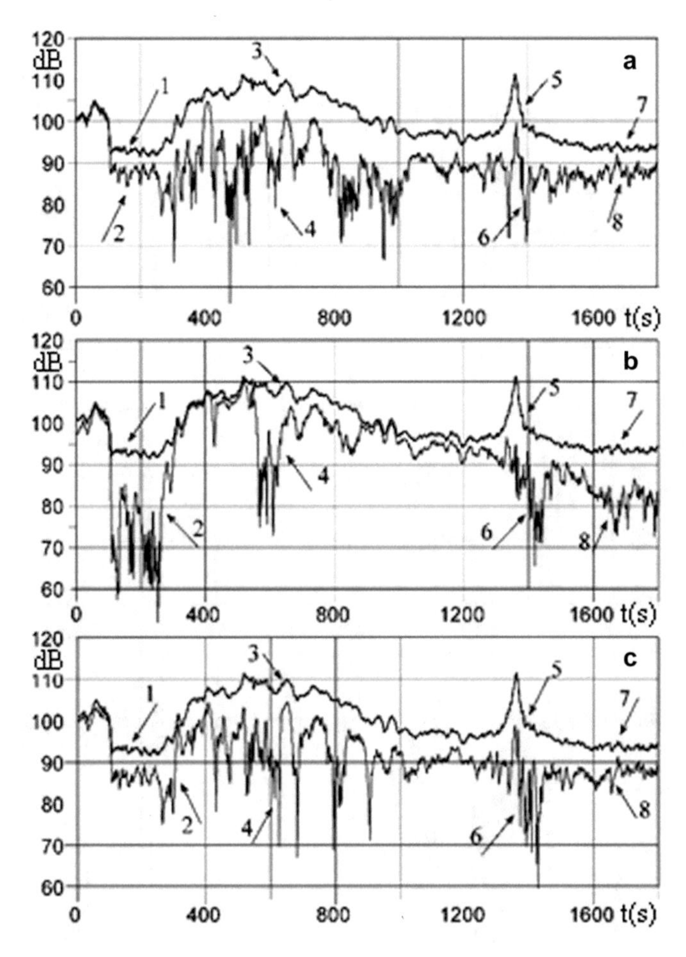

Fig. 5.7 Comparative passage characteristics; same conditions as in Fig. 5.5. Averaging: 10 s

grad p vector and the particle velocity vector V is $\pi/2$, in Figs. 5.6, 5.8 and 5.10 phase difference of $-90°$ corresponds to $180°$, but $+90°$ corresponds to $0°$ for particle velocity. Differential phase characteristics indicate that the background noise (1, 2 and 7, 8) travels in the $-x$ direction; on the y axis, from the $+y$ direction; and on the z axis, the noise travels from above, from surface-to-bottom.

Interval 3, 4 contains diving boat noise; interval 5, 6, the noise of a yacht that follows the diving boat, which is why the phase differences match along the x and y axes. Differential phase $\Delta\varphi_{pV_z}$ indicates the presence of the diving boat and the yacht at the time ($t = 900$ s on the x axis, $t = 600$ s on the y axis) when background noise dominates the x and y axes.

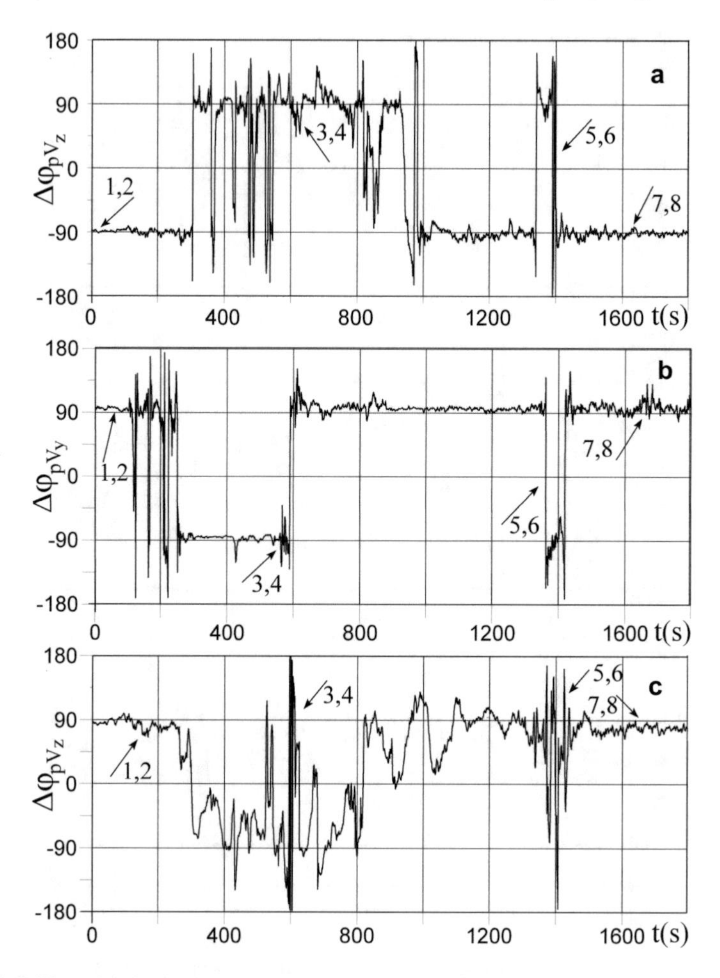

Fig. 5.8 Differential phase passage characteristics for Fig. 5.7. Averaging: 10 s

5.4 Vector-Phase Passive Acoustic Sonar

The proposed method of signal processing is fundamentally different to the processing in existing sonar systems, which use scalar acoustic variables. The SNR signal-to-noise ratio of receiving sonar systems built on scalar receivers (hydrophones) is given by the sonar equations:

$$\begin{aligned} &\text{for active sonars SNR} = \text{S L} - 2\text{TL} + \text{TS} - (\text{NL} - \text{DI}) \\ &\text{for passive sonars SNR} = \text{SL} - \text{TL} - (\text{NL} - \text{DI}), \end{aligned} \tag{5.6}$$

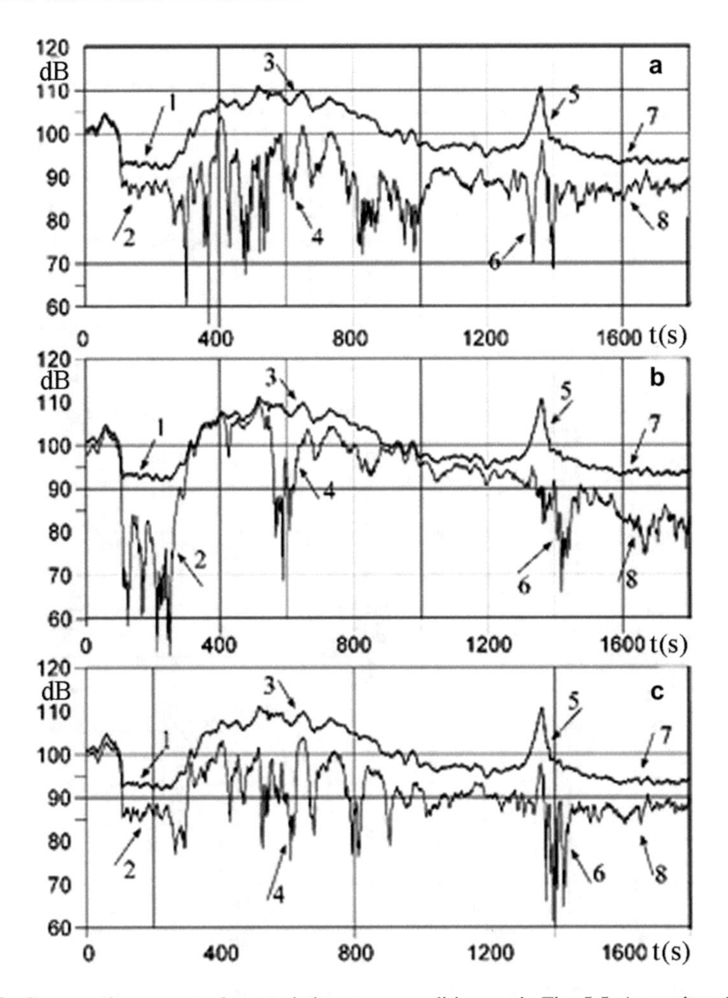

Fig. 5.9 Comparative passage characteristics; same conditions as in Fig. 5.5. Averaging: 15 s

where SNR is normalised signal-to-noise ratio at the output of the receiver (dB); SL is acoustic source level (dB) re 0.667×10^{-18} W/m^2 (this is a unit of sound intensity in a plane wave with an RMS sound pressure of 1 $\mu V/\text{Pa}$); TL is one-way transmission loss (dB) from the radiation source to the target or from the target to the receiver; TS is target strength (dB) for an active sonar, this describes how effectively the target reflects sound; NL is noise level in the receiver (dB) re 1 $\mu V/\text{Pa}$; DI is directivity index of the receiving antenna (dB). This variable shows to what degree the hydrophone antenna attenuates isotropic noise. The SNR estimation methodology for passive detection is determined by the DI parameter (directivity index of the receiving antenna), that is, the degree of coherence of useful signal in the antenna aperture on hydrophones spaced $l/2$; coherent interference noise is also amplified in the antenna aperture along with the signal.

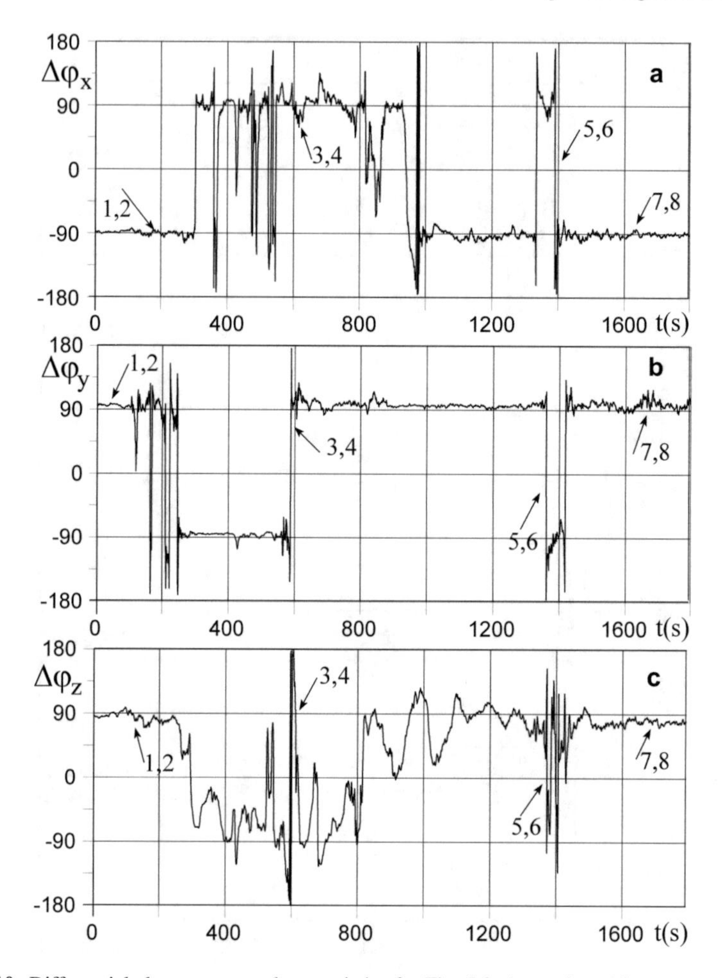

Fig. 5.10 Differential phase passage characteristics for Fig. 5.9. Averaging: 15 s

The proposed method of detection of acoustic sources differs from detection methods that use scalar antennas. As is known, an antenna's detection threshold is found by comparing signal strength and noise. A receiving system's range depends on its threshold signal power—the minimum power of useful signal at the input of the receiver at which the specified probabilities of missed detection and false alarm are achieved with a given accuracy. In a scalar field, this is found from the familiar sonar equation. But in vector acoustic fields, a weak signal can be completely compensated by anisotropic noise (interference) field, which frustrates comparisons of signal and noise strengths and opens up new possibilities for detection.

Fundamental properties of acoustic fields inherent to their vector nature can be used in real-life underwater sound applications. An individual combined receiver of a small wavelength size with multiplicative signal processing, being a correlation receiver, can measure time coherence and, crucially, its anisotropy at a given point in the acoustic field. This monograph discusses the possibility of constructing an algorithm for detecting weak signals in complex acoustic conditions of coherent interference noise and shallow water dynamic noise by harnessing the phenomenon of compensation of reciprocal energy fluxes and the statistical instability of differential phase relations associated with the compensation process [1, 13]. A weak signal is one whose power is comparable with that of interference noise; in this case, signal-to-interference ratio in the pressure channel shouldn't exceed 3 dB.

At the heart of a detection algorithm based on a specific physical phenomenon is the researcher's a priori knowledge of the criteria of useful signal associated with the phenomenon's properties. With compensation of reciprocal energy fluxes, the *a priori* information for the detection algorithm is the 'dip' in coherent power and the frequency coherence function relative to interference and noise at the frequency (in the frequency band) of the desired signal. Phase spectra at that frequency must either contain a steady 180° jump of phase difference with incomplete compensation or a random indefinite phase difference with full compensation in the same channels.

The algorithm must be superior to the passive detection algorithm for scalar receiving systems that comes down to a well-known procedure for estimating sample functions of a random process [17]:

$$\begin{aligned} r(t) &= s(t) + n(t), \quad T_i \le t \le T_j{:}\, H_1 \\ r(t) &= n(t), \quad T_i \le t \le T_j{:}\, H_0. \end{aligned} \tag{5.7}$$

In (5.7), $r(t)$ is a sample of the random process function; $s(t)$ is signal; $n(t)$ is noise; $T_i - T_j$ is observation time interval; H_1, H_0 are hypotheses about the presence and absence of a signal. The statistical nature of the detection process is based on the method of testing statistical hypotheses and parameter estimation. To circumvent difficulties arising from unknown signal amplitude and a priori probabilities to choose an acceptable false alarm probability, we use the Neyman–Pearson criterion. The detection threshold is chosen so as to maximise the probability of detection for the given probability of false alarm.

Evaluating sample random functions of a vector field given a priori statistical information greatly simplifies decision-making in the detection problem. The direction to the sound source is found from the formula $\alpha = \arctan S_{pV_y}(f,t)/S_{pV_x}(f,t)$ or by the jump of phase difference $\Delta\varphi_x(f,t)$, $\Delta\varphi_y(f,t)$ as the directivity pattern is rotated. Angular resolution of two closely spaced sources is accomplished using the Rayleigh criterion [1, 14].

5.4.1 Operating Principle of the Passive Sonar

Unlike the classical model of the detection device based on a square-law detector, in which detection amounts to a mere comparison of signal and noise strengths, we propose (in addition) a model of a detection device based on analysis of signal phase and degree of coherence using a phase detector and a coherent detector. The probabilistic statistical process of detection and decision-making complements the classical detection procedure (Neyman–Pearson criterion) by introducing a new statistical decision-making procedure that uses the Rayleigh criterion. Figure 5.11 is a schematic of the vector-phase sonar.

The operating principle of the sonar is based on the distinctive properties of the vector acoustic field: the compensation phenomenon, the consistent behaviour of differential phase relations at low SNR $\ll 1$ and a 180° phase jump when the target crosses the minimum of the dipole directivity pattern. These conditions are met whatever the field structure [14, 15].

A flowchart of the sonar is shown in Fig. 5.11; this underpins the detection method and 'target–no target' decisions.

The sonar is based on an individual combined receiver. Bearing to the target is estimated by electronically rotating and sweeping the x and y directivity patterns about the z axis through an angle $\pm\, \Delta a$ about the direction to the target. The noisy target is localised on the extended perpendicular to the x axis if the x directivity pattern sweeps about its minimum or on the extended perpendicular to the y axis if the y directivity pattern sweeps about the y pattern minimum. The perpendiculars correspond to the minima of the x and y patterns. Good-quality vector receivers have sensitivity relation coefficients about 26–29 dB in angular sectors no wider than 3°, which ensures high-accuracy direction finding at the minimum of the directivity pattern. The minimum of the x channel pattern matches the maximum of the y channel,

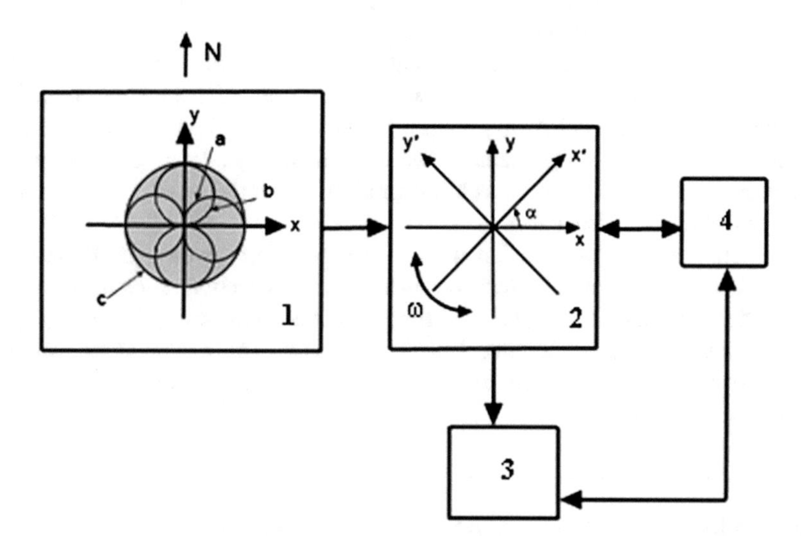

Fig. 5.11 Flowchart of the vector-phase sonar. *Legend* 1-angular sensitivity characteristics of horizontal channels x and y: a, b-channels x and y of the vector receiver, respectively; c–acoustic pressure channel; 2-electronic directivity pattern rotator; ω-angular rotation ('sweep') velocity of the combined receiver's directivity pattern; 3-human operator; 4-Fourier and Hilbert processors, coherent power detector, phase detector

which is important to signal processing. The main object of location is the 180° jump of phase difference $\Delta\varphi_x(f_0, \alpha)$ or $\Delta\varphi_y(f_0, \alpha)$ and the peaks of their derivatives $\Delta\dot{\varphi}_x(f_0, \alpha) = \frac{\mathrm{d}(\Delta\varphi_x(f_0,\alpha))}{\mathrm{d}\alpha}$ or $\Delta\dot{\varphi}_y(f_0\alpha) = \frac{\mathrm{d}(\Delta\varphi_y(f_0,\alpha))}{\mathrm{d}\alpha}$. At the inflection point of phase difference curves, derivatives $\Delta\dot{\varphi}_x(f_0, \alpha)$ α) and $\Delta\dot{\varphi}_y(f_0, \alpha)$ α) have sharp positive or negative peaks associated with phase 'reversal' from 0° to 180° or from 180° to 0°. The other primary approach is to use compensation of reciprocal energy fluxes in the detection methodology—primarily the spectral dips at the weak signal frequency and instability of phase difference at that frequency. The operating principle of the sonar is based on the author's patents [14, 15].

Figures 5.12, 5.13, 5.14 and 5.15 illustrate the behaviour of phase differences $\Delta\varphi_{x,y}(f_0, t)$, coherence functions $\mathrm{Re}\,\Gamma_{x,y}(f_0, t)$ and their time derivatives $\Delta\dot{\varphi}_{x,y}(f_o, t)$, $\mathrm{Re}\,\dot{\Gamma}_{x,y}(f_o, t)$ when the signal from an underwater target crosses the

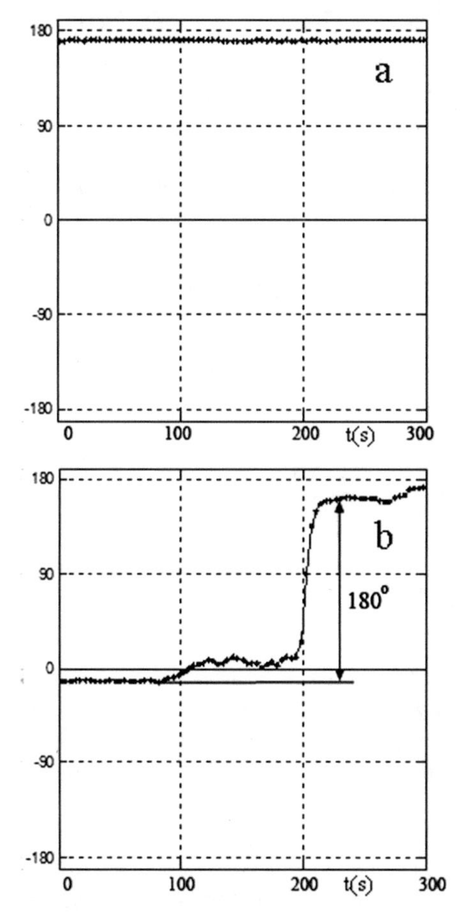

Fig. 5.12 Phase differences: **a** $\Delta\varphi_x(f_o, t)$, **b** $\Delta\varphi_y(f_o, t)$. Averaging time: 8 s. Frequency: 400 ± 5 Hz

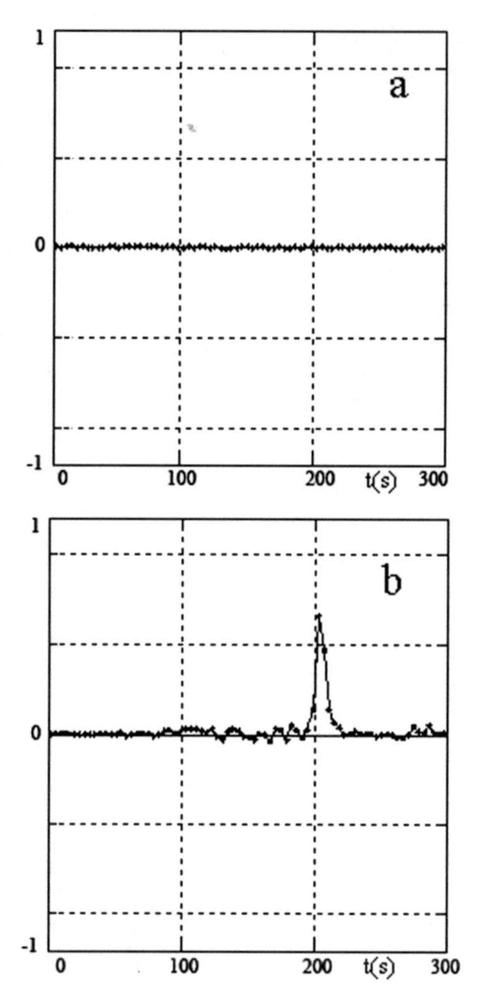

Fig. 5.13 Time derivatives of the phase differences in Fig. 5.12 as functions of time: **a** $\Delta\dot{\varphi}_x(f_0, t)$, **b** $\Delta\dot{\varphi}_y(f_o, t)$. Averaging time: 8 s

minimum of the y channel directivity pattern. Since the target is at the maximum of the x channel directivity pattern, variables $\Delta\varphi_x(f_o, t)$, $\Delta\dot{\varphi}_x(f_o, t)$, $\mathrm{Re}\,\Gamma_x(f_o, t)$ and $\mathrm{Re}\dot{\Gamma}_x(f_o, t)$ are featureless in time. However, these variables do have features in the y channel. Phase differences $\Delta\varphi_y(f_0, t)$ and coherence function $\mathrm{Re}\,\Gamma_y(f_0, t)$ experience a jump: $\Delta\varphi_y(f_0, t)$ by 180°, $\mathrm{Re}\,\Gamma_y(f_0, t)$ from -1.0 to $+1.0$; at the transition point at $t \approx 200$ s their derivatives $\Delta\dot{\varphi}_y(f_o, t)$, $\mathrm{Re}\,\dot{\Gamma}_y(f_o, t)$ too experience a jump, which is clearly visible against the otherwise zero background.

In the vector acoustic field [1, 13–15] with compensation of reciprocal energy fluxes, statistical characteristics of parameters $SpV_i(f)$, $\gamma_i^2(f)$, $\Delta\varphi_i(t)$ may indicate

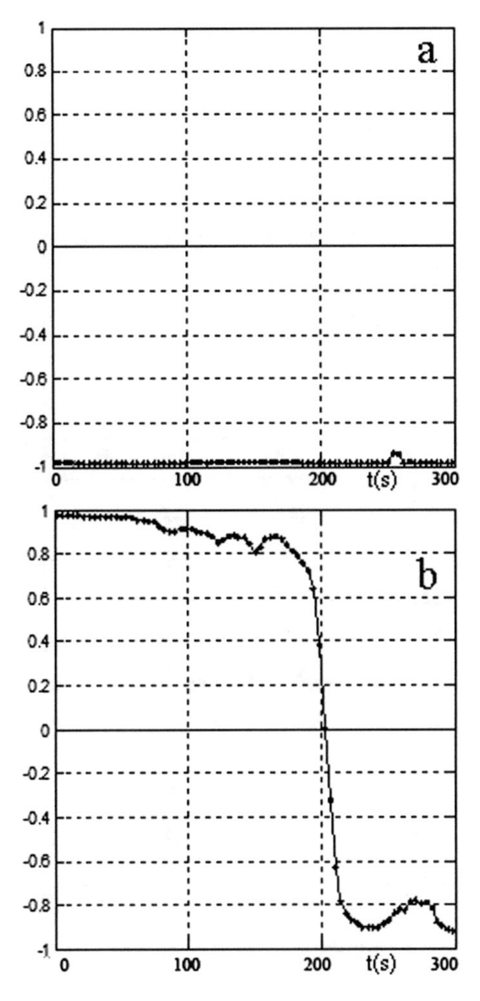

Fig. 5.14 Time dependence: **a** $-\text{Re}\,\Gamma_x(f_0, t)$; **b** $\text{Re}\,\Gamma_x(f_0, t)$). Averaging time: 8 s. Same realisation as in Figs. 5.12 and 5.13

the presence of a signal in a mixture of coherent interference and underwater ambient noise, with the added benefit of a significant gain in signal-to-noise ratio.

Therefore, the fundamental properties of the vector acoustic field can be exploited to create a novel detection device: the vector-phase sonar.

The optimum detection device based on an individual combined receiver of small wavelength dimensions consists of a coherent power detector and a phase detector. Statistical data processing is based on the Fourier and Hilbert transforms. Hypotheses H_1 and H_0 are tested on simultaneous match of a priori data, namely dips in the relevant components of coherent power, frequency coherence function and the relevant differential phase relations.

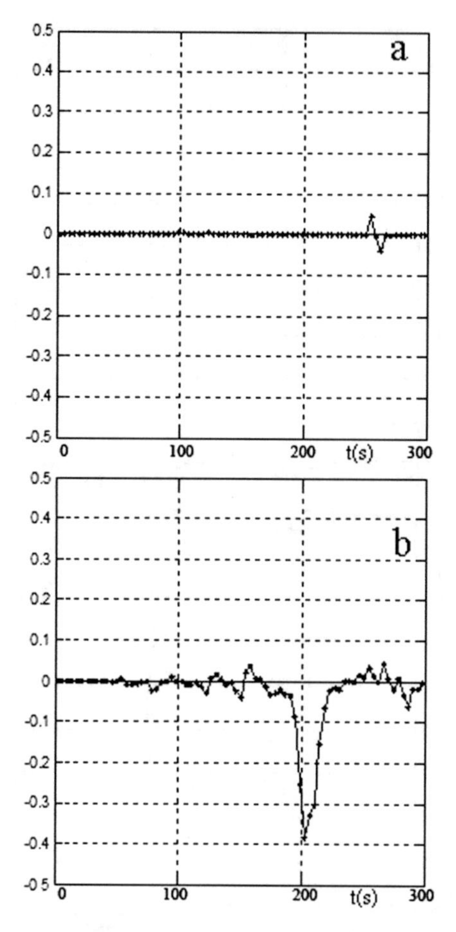

Fig. 5.15 Time dependence of derivatives: **a** Re $\dot{\Gamma}_x(f_o, t)$, **b** Re $\dot{\Gamma}_y(f_o, t)$. Averaging time: 8 s

5.4.2 *Sonar Data Processing Sequence*

A human operator (point 3 in the caption of Fig. 5.11) uses an electronic rotator to command the directivity pattern to rotate at an angular velocity of $\omega = \dot{\alpha}$, (where $\dot{\alpha} = \frac{d\alpha}{dt}$) using the Cartesian rotation matrix (Fig. 5.14)

$$\begin{aligned} u_x(\alpha) &= u_x \cos\omega t + u_y \sin\omega t \\ u_y(\alpha) &= -u_x \sin\omega t + u_y \cos\omega t, \end{aligned} \tag{5.8}$$

where u_x and u_y are electrical signals when x and y axes point in the direction N and $a = 0$ (Fig. 5.11), $u_x(\alpha)$ and $u_y(\alpha)$ are electrical signals when axes x and y turn through the angle a.

The computer performs the following real-time mathematical operations while in rotation mode:

$$\begin{aligned}
&1.\,\mathrm{Re}\,I_{x,y}(t,\,f_0,\alpha(t)),\,\mathrm{Im}\,I_{x,y}(t,\,f_0,\alpha(t))\\
&\mathrm{Re}\,\Gamma_{x,y}(t,\,f_0,\alpha(t)),\,\mathrm{Im}\,\Gamma_{x,y}(t,\,f_0,\alpha(t))\\
&2.\,\varphi_{x,y}(t,\,f_0,\alpha(t)),\quad \dot{\varphi}_{x,y}(t,\,f_0,\alpha(t))
\end{aligned} \tag{5.9}$$

Once it detects a $\pm 180°$ 'reversal' of the phase difference, which indicates the presence of a local sound source, the system determines the angle α_0 of this 'reversal'. Next, the x and y axes are set to 'sweep' relative to α_0, that is, $\alpha_0 \pm \Delta\alpha$, and derivatives $\dot{\varphi}_{xy}(\Delta\alpha)$ are computed for Δa varying within $\pm 5°$. If the 'sweeping' is about the minimum of the x channel, the maximum of the y channel points at the target and its derivative $\dot{\varphi}_y(\Delta\alpha) = 0$. Derivative $\dot{\varphi}_x(\Delta\alpha)$ has its maximum at α_0. In a real situation, the angle $\alpha_0(t)$ depends on time due to movement of the target; as a result, derivative $\dot{\varphi}_x(\Delta\alpha)$ also shifts in time. As it does, the derivative $\dot{\varphi}_x(\Delta\alpha)$ will track the movement of $\alpha_0(t)$; accordingly, the peak of the derivative will point at the target. At the same time, derivative $\dot{\varphi}_y(\Delta\alpha)$ must be zero because the maximum of the y pattern points precisely at the target. Concurrently, sonogram $S_{pV_y}(f, t)$ is plotted for the y channel and energy characteristics of the target are monitored. Spectra $S_{p^2}(f)$, $S_{p,V_{x,y}}(f)$ at the target frequency are used to determine whether compensation occurs or not. Target validation decision is made after several iterations of varying the angle $\pm\,\Delta\alpha$. If the probability of $\pm\,180°$ jumps of phase difference and the corresponding peaks of derivatives tends to unity, the target is validated. Since the locally detected target is set against a mixture of anisotropic dynamic noise and interference noise, it is assumed that variations of dynamic noise within the angle $\Delta\alpha \approx \pm$ (3–5°) are insignificant and there is one local target in a (6–10°) sector. A result of phase location is shown for illustration (Fig. 5.16) [14].

5.4.3 Fourier and Hilbert Signal Processing Sequences

Figures 5.17 and 5.18 show block diagrams of Fourier and Hilbert signal processing schemes respectively.

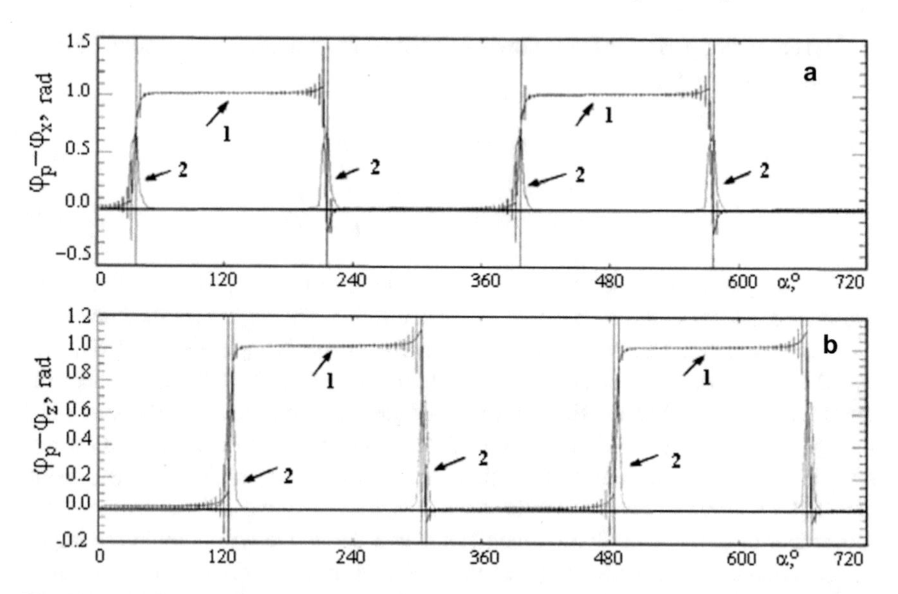

Fig. 5.16 A real result of passive location of an underwater source. Source depth ≈ 60 m. Combined receiver depth: 150 m. **a** Jumps of phase difference $\Delta\varphi_x(\alpha_0, t_0)$; **b** of $\Delta\varphi_x(\alpha_0, t_0)$. 1—$\Delta\varphi_x(\alpha_0, t_0)$ and $\Delta\varphi_y(\alpha_0, t_0)$ curves, 2—their derivatives. The vertical lines on the curves correspond to σx and σy. Rotation angle α is in degrees, and phase differences in radians

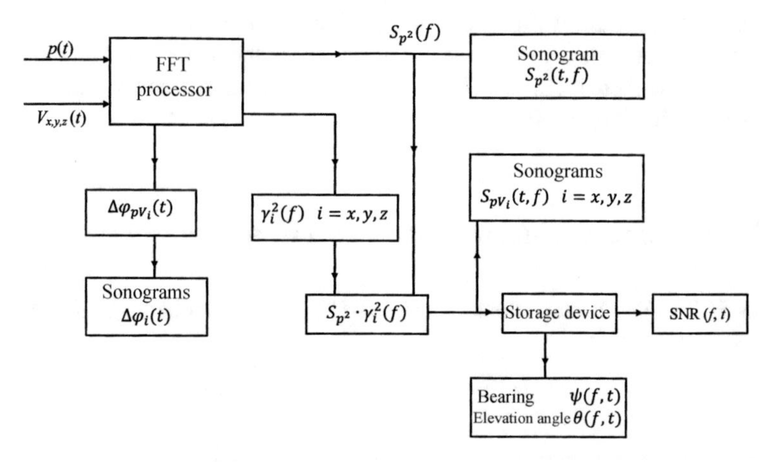

Fig. 5.17 Flowchart of mathematical processing of vector characteristics of signal and underwater ambient noise using FFT

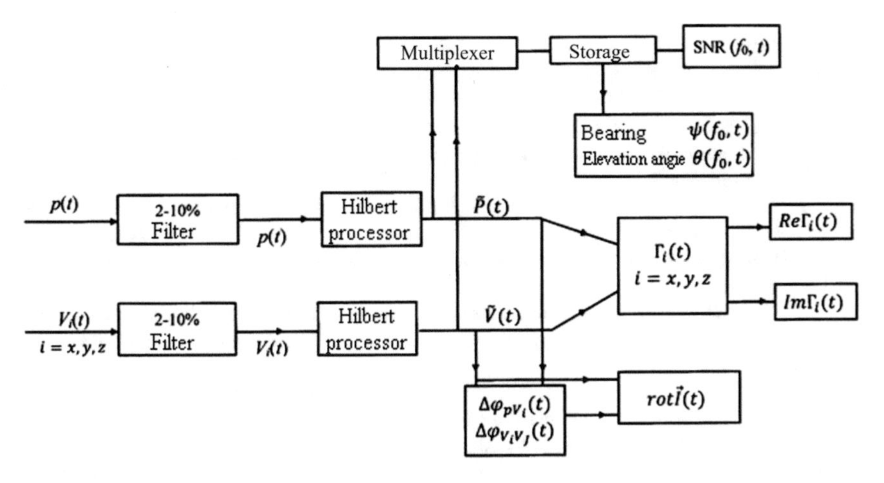

Fig. 5.18 Flowchart of mathematical processing of vector characteristics of signal and underwater ambient noise using the Hilbert transform

5.5 Conclusions

We have outlined the theoretical foundations of combined reception and analysed experimental data. Noise immunity gain of SNR(PV) compared with SNR(P^2) can be as high as ~ 30 dB in anisotropic field. We have formulated the principal phenomena found in the vector acoustic field of the real ocean which should be applied in observation of weak signal in diffuse, partially coherent and coherent acoustic interference noise fields. The presented material is original and can be used in fundamental and applied research of the real ocean.

References

1. V.A. Shchurov, *Vector Acoustics of the Ocean* (Vladivostok, Dalnauka, 2003). (in Russian)
2. V.A. Shchurov, A.V. Shchurov, Noise immunity of the hydroacoustic combined receiver. Akusticheskij Zhurnal (Acoust. Phys.) **48**(1), 110–119 (2002). (in Russian)
3. M.D. Smaryshev, On noise immunity of the combined hydroacoustic receiver. Akusticheskij Zhurnal (Acoust. Phys.) **51**(4), 558–559 (2005). (in Russian)
4. V.A. Shchurov. A reply to M. D. Smaryshev's letter. Akusticheskij Zhurnal (Acoust. Phys.) **51**(4), 560–561. (in Russian)
5. M.D. Smaryshev, E.L. Shenderov, Noise immunity of flat antennas in anisotropic interference field. Akusticheskij Zhurnal (Acoust. Phys.) **31**(4), 502–506 (1985). (in Russian)
6. A.V. Pesotsky, M.D. Smaryshev, Comparative evaluation of efficiency of receiving antennas consisting of combined receivers in a free field and near a flat screen. Akusticheskij Zhurnal (Acoust. Phys.) **35**(3), 495–498 (1989). (in Russian)
7. E.L. Shenderov, On noise immunity of antennas consisting of sound pressure receivers and particle velocity receivers. Gidroakustika (Hydroacoustics) **3**, 24–40 (2002). (in Russian)

8. V.A. Gordienko, E.L. Gordienko, N.V. Krasnopistsev, V.N. Nekrasov, Noise immunity of hydroacoustic receiving systems that register acoustic power flux. Akusticheskij Zhurnal (Acoust. Phys.) **54**(5), 774–785 (2008). (in Russian)
9. D.V. Zlobin, B.A. Kasatkin., S.B. Kasatkin, G.V. Kosarev, Some results of investigations of scalar-vector sound fields in the infrasonic range. Gidroakustika (Hydroacoustics) **31**(3), 65–78 (2017). (in Russian)
10. B.A. Kasatkin, S.B. Kasatkin, Experimental evaluation of noise immunity of a combined receiver in the infrasonic range. Underwater Invest. Robot. **1**(27), 38–47 (2019). (in Russian)
11. V.P. Dzyuba, *Scalar-Vector Methods of Theoretical Acoustics* (Dalnauka, Vladivostok, 2006). (in Russian)
12. G.M. Glebova, G.A. Zhbankov, I.A. Seleznev, Analysis of signal detection characteristics of a vector-scalar antenna receiving system. Gidroakustika (Hydroacoustics) **19**(1), 68–78 (2014). (in Russian)
13. V.A. Shchurov, V.P. Kuleshov, E.S. Tkachenko, E.N. Ivanov, Determinants of compensation of reciprocal energy fluxes in the acoustic fields of the ocean. Akusticheskij Zhurnal (Acoust. Phys.) **56**(6), 835–843 (2010). (in Russian)
14. V.A. Shchurov, E.N. Ivanov, I.A. Ivanov, Method of determining bearing of noisy object: pat. 2444747 C1 RU. No. 2010126808; filed 30.06.10; publ. 10.03.12, Bull. 7
15. V.A. Shchurov, E.N. Ivanov, G.F. Ivanova, Multichannel digital combined hydroacoustic system: pat. 82972 U1 RU, no. 2008152703; filed 30.12.08; publ. 10.05.09, Bull. 13
16. V.A. Shchurov, E.N. Ivanov, S.G. Shcheglov, A.V. Cherkasov, An underwater glider for monitoring acoustic vector fields: pat. 106880 U1 RU, no. 2011108806; filed 09.03.11; publ. 27.07.11, Bull. 21
17. W.S. Burdic, *Underwater Acoustic System Analysis*, Russian. (Sudostroenie, Leningrad, 1988)

Chapter 6
Vector-Phase Experimental Technique, Expeditions, Conferences

This chapter is dedicated to the history of research and establishment of the vector-phase method at the Pacific Oceanological Institute (Ocean Acoustic Noise Laboratory) (1980–2019). The photographs depict the experimental technique over the years, participation in international conferences and international relations (Figs. 6.1, 6.2, 6.3, 6.4, 6.5, 6.6, 6.7, 6.8, 6.9, 6.10, 6.11, 6.12, 6.13, 6.14, 6.15, 6.16, 6.17, 6.18, 6.19, 6.20, 6.21, 6.22, 6.23, 6.24, 6.25, 6.26, 6.27, 6.28, 6.29, 6.30, 6.31, 6.32, 6.33, 6.34, 6.35, 6.36 and 6.37).

6.1 Field Research

6.1.1 R/V Callisto Cruise. Kuril–Kamchatka Chain. May–June 1979

The first ever vector-phase studies of anisotropy of low-frequency underwater acoustic noise in the Pacific Ocean. The experiment inaugurated vector-phase measurements in underwater acoustics.

6.1.2 Northwestern and Central Pacific. R/V Balkhash Cruise. 1983

Experiment with a combined vertical antenna at depths of 200–1000 m. Investigation of anisotropy of low-frequency underwater acoustic noise in the ocean, man-made interference noise and noise immunity of receiving systems in the real ocean environment. *Acquired data on extremely long-range propagation of low-frequency sound.*

V. A. Shchurov, *Movement of Acoustic Energy in the Ocean*,
https://doi.org/10.1007/978-981-19-1300-6_6

Fig. 6.1 President of the Russian Academy of Sciences Yu. S. Osipov and Vice-President N. P. Laverov in V. A. Shchurov's laboratory, which at the time (1995–1998) conducted research under contract with Harbin Engineering University

6.1.3 Northwestern and Central Pacific; Indian Ocean. R/V Akademik Vinogradov. 1990

Investigation of vector acoustic properties of the deep open ocean using a free-drifting combined telemetry system. The expedition followed a route from near the Kamchatka Peninsula to southern latitudes of the Indian Ocean. The object of research was low-frequency noise and signal. Frequencies ranged from 5 to 1000 Hz, and depths from 150 to 1000 m. The photographs depict expedition work in the Indian Ocean.

6.2 Shallow Water Acoustic Research Coastal Expeditions

Investigations of acoustic processes in shallow water were carried out in the Peter the Great Bay at depths of ~ 30 to ~ 120 m. The most exciting results were detection of vortices of the acoustic intensity vector, investigation of directionality of underwater ambient noise and the phenomenon of compensation of reciprocal energy fluxes. Research cruises helped to field-test vector-phase reception techniques and develop acoustic equipment for deep open ocean research.

In 1980 this became the building of the Ocean Acoustic Noise Laboratory by order of V.I. Ilichev, then director of the Pacific Oceanological Institute. It was here that Professor Ilichev would frequently do experimental work as an ordinary researcher (1980–1994).

Fig. 6.2 Bottom-mounted combined low-frequency four-component acoustic receiving system co-designed by Moscow State University and the Pacific Oceanological Institute, seen here aboard R/V *Callisto*, Iturup Island, Barkhatnaya Cove. Pictured in the centre is V. A. Shchurov (POI) and members of the Acoustics Department, Faculty of Physics, MSU. Left to right: A. Slutskov, S. Ilyin, F. Toporovsky

Fig. 6.3 Experiment aboard R/V *Callisto*. On board the sailing/motor yacht *Blues*. Barkhatnaya Cove. Recording noise and signal from the combined bottom-mounted system. S. A. Ilyin, V. A. Shchurov, June 1979

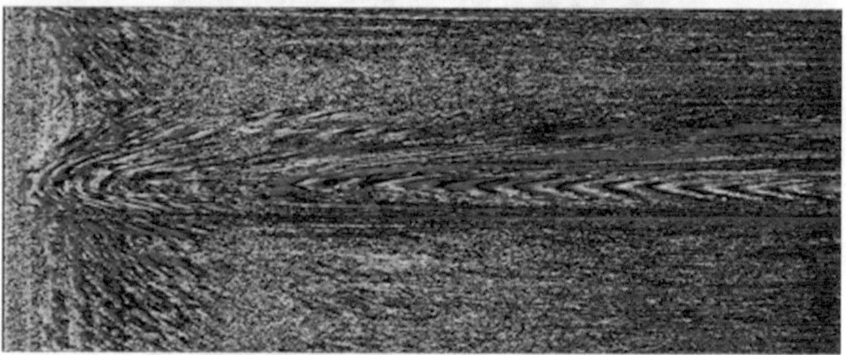

Fig. 6.4 Launching the antenna from R/V *Balkhash*. Beaufort scale 3 seas. Northwestern Pacific. 1983. The bottom panel shows a phase sonogram of a broadband signal

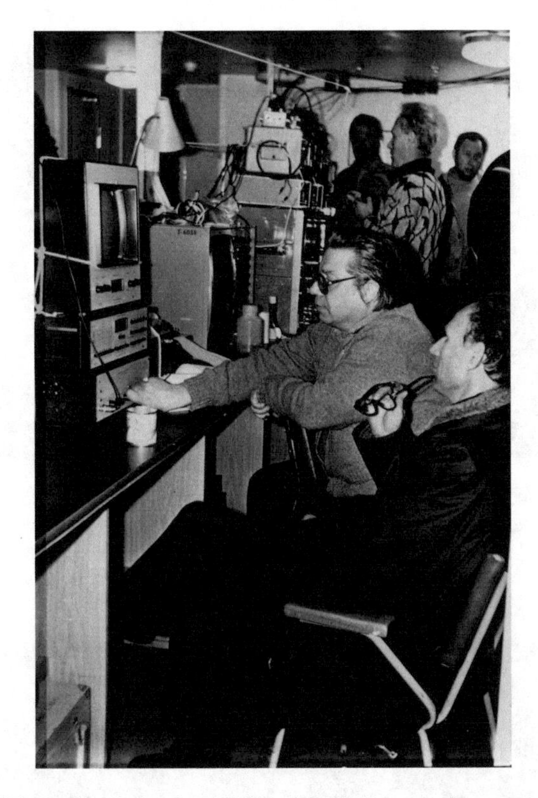

Fig. 6.5 Aboard R/V *Balkhash*. Academy Fellow V. I. Ilichev controls antenna directivity pattern

Fig. 6.6 R/V *Balkhash*. Analogue recording stack. V. A. Shchurov, 1983

Fig. 6.7 R/V *Balkhash*. V. I. Ilyichev, V. A. Shchurov. Analysing low-frequency radiation from V. A. Akulichev aboard R/V *A. Vinogradov*, 1983

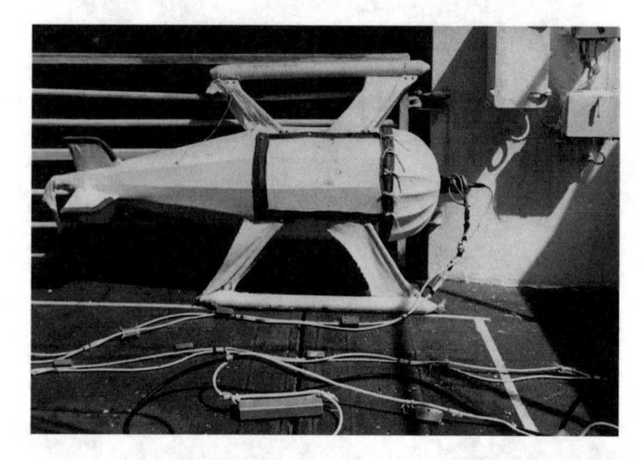

Fig. 6.8 Neutrally buoyant module. A joint design by the Pacific Oceanological Institute, KNIIGP (Kyiv Research Institute of Hydrological Instruments, Kyiv) and Shtorm Design Bureau (Kyiv Polytechnic Institute, Kyiv), NPO Leninets, Leningrad

Fig. 6.9 Pre-launch conference. Left to right: N. K. Voronin, N. G. Klimenok (KNIIGP), V. A. Shchurov (POI), V. P. Rudnichenko, Y. A. Khvorostov, L. F. Shikov, V. A. Komarov (KNIIGP)

Fig. 6.10 System being assembled on board R/V *Akademik Vinogradov*

Fig. 6.11 Launching from the R/V's ramp. Left to right: A. S. Fedoseenkov, S. P. Lisitsa, V. A. Shchurov, V. A. Meleshko, V. I. Kovalenko

Fig. 6.12 The module descends to a depth of 1000 m

Fig. 6.13 The harsh southern latitudes of the Indian Ocean. RF transmitter buoy of the telemetry system bobs in Beaufort scale 3 seas. The white dots on the surface are plastic floats of the cable line descending into the depths of the ocean. Two modules are underneath, at depths of 500 and 1000 m

Fig. 6.14 R/V *Akademik Lavrentyev* by the rocky shores of Kamchatka, Russkaya Cove, 1987. Left to right: V. A. Meleshko, V. A. Shchurov, Yu. A. Khvorostov, V. A. Komarov, V. I. Kovalenko, S. P. Lisitsa, at the top: A. N. Fadeev. Northern latitudes of the Pacific, our lab room

Fig. 6.15 Ocean Acoustic Noise Laboratory building, Vityaz Cove. 2019

Fig. 6.16 Academy Fellow V. I. Ilichev played an invaluable role in establishing and developing the vector-phase approach in the Russian Far East, and especially its use in applied research. Pictured: V. I. Ilichev and V. A. Shchurov, Vityaz Cove, 1986

Fig. 6.17 Discussing experimental data. On the left: academy fellow V. I. Ilichev and V. A. Shchurov; on the right: researchers of the Ocean Acoustic Noise Laboratory, Vityaz Cove, 1986

Fig. 6.18 R/V *Gidronavt*, the workhorse of the shallow sea, seen here launching the system at the shelf break. $H = 120$ m. Astern next to the combined receiving system module are S. G. Shcheglov, L. F. Shikov, V. A. Shchurov and Yu. A. Khvorostov (1982)

Fig. 6.19 Researcher Elena Tkachenko processes data on a four-channel analogue-to-digital stack, 1982

Fig. 6.20 The moment of truth: launching the combined module. Peter the Great Bay, 2015

Fig. 6.21 A 16-channel module capable of resolving the structure of an acoustic vortex. A. S. Lyashkov. Vityaz Cove, 2014

Fig. 6.22 Preparing modules for the launch. S. Shcheglov, D. Strobykin, M. Lebedev, 2014

Fig. 6.23 Preparing the cable line. The cable line can be 300–1500 m long. Schultz Point Marine Experimental Base, 2014

Fig. 6.24 Aboard the sailing yacht *Svetlana*. Recording noise in the Peter the Great Bay. V. A. Shchurov, 2015

Fig. 6.25 V. A. Shchurov and B. A. Kasatkin. Discussing a joint experiment. 2014

6.3 International Relations

6.3.1 People's Republic of China

Photographic illustrations of the collaboration with scientists from the People's Republic of China are shown below in Figs. 6.26, 6.27, 6.28 and 6.29.

Fig. 6.26 Signing of a contract between the Pacific Oceanological Institute and Harbin Engineering University. Pictured: academy fellow Yang Shi Ye and V. A. Shchurov. Harbin, 1996

Fig. 6.27 Celebrating the completion of experiments at Songhua Lake. Pictured on the left of V. A. Shchurov is academy fellow Yang Shi Ye, on the right interpreter M. V. Kuyanova and Sang Enfang. Harbin, 1997

Fig. 6.28 International acoustic conference, Harbin, 1997. First row, left to right: second—Sang Enfang (China), third—Yang Shi Ye (China), fourth—M. Zakharia (France), fifth—L. Bjørnø (Denmark), seventh—R. Spindel (USA). Second row: behind R. Spindel—V. Shchurov, left to him—W. Kuperman and then J. Lynch (USA)

Fig. 6.29 Chinese exchange students in the Far Eastern Federal University. Professors V. I. Korochentsev and V. A. Shchurov after lectures among acoustics students from Harbin Engineering University. Far Eastern Federal University. Vladivostok. Russky Island, 2017

6.3.2 USA and UK

Photographic illustrations of the collaboration with USA and UK scientists are shown below in Figs. 6.30, 6.31, 6.32 and 6.33.

Fig. 6.30 First international conference on global acoustic ocean observing. Scripps Institution. San Diego, California, USA, 8–9 June 1992. Photographed by V. A. Akulichev

Fig. 6.31 Vector-phase method acoustics conference. Applied Physics Laboratory, University of Washington. V. A. Shchurov, R. C. Spindel, G. L. D'Spain. Seattle, USA, 1999

Fig. 6.32 12:00 noon sharp Greenwich mean time on the Big Ben. London, 1997

Fig. 6.33 Natural physical sources of underwater sound symposium. University of Cambridge. At M. S. Longuest-Higgins's country home. Photo by H. Medwin. England, 1990

6.4 Promising Areas

6.4.1 Low-Frequency Acoustic Intensity Interferometer. Investigation of Coherent Properties of Acoustic Intensity in Spatially Distanced Points of Acoustic Field Using the Correlation Theory of Coherence

See Fig. 6.34.

6.4.2 Vector Geophone. Acoustic Studies at the Water–Bottom Interface of the Shallow Water Waveguide

Three-component vector geophones are used to investigate vector acoustic field in water near the bottom, on its surface and below the bottom surface.

Ocean Acoustic Noise Laboratory pioneered vector-phase studies in the Pacific Ocean (1979). Its scientists and engineers continue to perfect the technology and methodology of the vector-phase method. The Pacific Oceanological Institute of the USSR (and later Russian) Academy of Sciences has developed and implemented vector-phase techniques for the real ocean and coastal areas. Its research has significantly contributed to modern fundamental underwater physical acoustics.

Fig. 6.34 Eight-channel combined receiving module in its housing. A receiving component of the acoustic intensity interferometer (a counterpart of the Young–Rayleigh optical interferometer)

Fig. 6.35 Three-component vector geophone

Fig. 6.36 Academy fellow G. I. Dolgikh and Prof. V. A. Shchurov discussing a new experiment-ready geophone. Vityaz Cove. 2019

Fig. 6.37 Ocean Acoustic Noise Laboratory team. Left to right: A. S. Lyashkov, S. G. Shcheglov, L. F. Shikov, V. A. Shchurov, E. S. Tkachenko, V. P. Kuleshov. 2012

Appendix A
Monochromatic Acoustic Vector Field. Fundamental Relationships

A.1 Introduction

The fundamental equations of acoustics connect scalar and vector acoustic variables. Scalar parameters include: sound potential, acoustic pressure and density of the continuous medium. Vector parameters include linear quantities: particle displacement, velocity and acceleration, and acoustic field intensity, a second-order parameter. The existence of sound potential of the particle velocity vector indicates that the particle velocity vector field is potential and vortex-free, which is not true for the energy flux density (intensity) vector. Acoustics, especially underwater, has favoured acoustic pressure measurements, since acoustic pressure is much more straightforward to measure in a real medium than particle velocity, and in a simplified form velocity can always be expressed through pressure as $V = p/\rho c$ (where ρ is medium density and c is the sound speed in the medium), which is why underwater acoustics has mostly focused on scalar characteristics of the acoustic field. However, analysis of experimental data shows that direct measurements of particle velocity are often necessary in order to fully describe the acoustic field. In mathematical processing of experimental data, we deal with narrowband signal ($\Delta\omega/\omega_0 < 1$, where ω_0 is the central frequency of the signal being processed). The acoustic field theory outlined in this monograph is based on the notion of harmonic (monochromatic) signal, which is sufficient to describe the phenomena being discussed here.

A.1.1 Complex Description of Harmonic Vector Acoustic Fields

We will consider the wavefield to be random, stationary and ergodic, and the signals harmonic (monochromatic). We will use harmonic notation for mathematical relations of acoustic pressure $p(t)$ and the particle velocity vector $\boldsymbol{V}(t)$.

V. A. Shchurov, *Movement of Acoustic Energy in the Ocean*,
https://doi.org/10.1007/978-981-19-1300-6

As is known, harmonic functions of the form

$$p_1 = p_0 \cos(\omega t - \varphi_0)$$
$$p_2 = p_0 \sin(\omega t - \varphi_0) \tag{A.1}$$

each solve the Helmholtz equation

$$\Delta p + k^2 p = 0 \tag{A.2}$$

where $\Delta = \nabla \cdot \nabla = \nabla^2 \stackrel{\text{def}}{=} \text{div grad}$ is the differential Laplace operator; p_0 is acoustic pressure amplitude, ω is angular frequency; t is time, φ_0 is the initial phase and $k = \omega/c$ is the wave number.

A complex linear combination of solutions (A.1) of the form

$$\tilde{p} = p_1 - ip_2 = p_0 e^{i\varphi} e^{-i\omega t} \tag{A.3}$$

also solves the Helmholtz equation. With the wave written this way, the harmonic dependence on time is described by the factor $e^{-i\omega t}$, which is independent of coordinates. The expression $p = p_0 e^{i\varphi}$ is called the complex amplitude of fluctuations; it depends only on the coordinates and determines the amplitude p_0 and phase φ of fluctuations of the medium at various points. The Helmholtz equation is satisfied by the full solution (A.3) as well as its complex amplitude $p = p_0 e^{i\varphi}$. Real amplitude p_0 does not solve the Helmholtz equation. This notation greatly simplifies operations such as differentiation and integration. Differentiation of the complex wave with respect to time is done by multiplying by $-i\omega$, and integration by dividing by $-i\omega$. We will introduce for $\tilde{p} = p_1 - i\,p_2 = pe^{-i\omega t}$ its complex conjugate

$$\tilde{p}^* = p_1 + i\,p_2 = p_0 e^{-i\varphi} e^{i\omega t} = p^* e^{i\omega t} \tag{A.4}$$

The change of the sign from $-i$ to $+i$ can be interpreted to mean that the latter combination corresponds to a negative frequency if in (A.4) ω were to be replaced with $-\omega$. If so, in operations with negative frequencies all complex quantities must also be replaced with their conjugates, and differentiation and integration with multiplication and division by $i\omega$. Complex particle velocity in the complex acoustic wave $\tilde{p} = p_0 e^{-i(\omega t - \varphi)}$ can be found by integrating the Euler formula

$$V = -\frac{1}{\rho} \nabla \int_{t_0}^{t} p\,\mathrm{d}t \tag{A.5}$$

$$
\begin{aligned}
\tilde{V} &= \frac{1}{i\rho\omega}\nabla\tilde{p} = \frac{1}{i\rho\omega}\left(\nabla p' \cdot e^{i\varphi} + i\nabla\varepsilon \cdot \nabla p' e^{i\varphi}\right) \\
&= \frac{1}{i\rho\omega}\left(\nabla \ln\, p' + i\nabla\varphi\right)p
\end{aligned} \tag{A.6}
$$

From (A.6) it follows that a particle in the harmonic wave moves in a plane; the particle's velocity is parallel to ∇p and $\nabla\varphi$, and its components along these vectors are a quarter of the period out of phase. Motion can be different at different points in the plane, resulting in an elliptical trajectory of the end of the velocity vector.

Complex velocity can be expressed through conjugate pressure by the formula

$$
\tilde{V} = \frac{1}{i\rho\omega}\nabla p^* = \frac{1}{-i\rho\omega}\left(\nabla \ln\, p' - i\nabla\varepsilon\right)p^*, \tag{A.7}
$$

and a plane wave travelling to the right can be written as

$$
p = p_0 e^{-i(\omega t + kx)}. \tag{A.8}
$$

Negative frequencies occur in the expansion of waves into Fourier integrals in the interval from $-\infty$ to $+\infty$ when the Fourier expansion is over complex exponents rather than over trigonometric functions. Taking the real part gives the same result regardless of the frequency sign: waves that differ only in the frequency sign are one and the same physical object despite the different mathematical notation. Complex notation is a convenient mathematical formalism. Complex solutions of the Helmholtz equation have no physical meaning—only their real parts do.

A.1.2 Plane and Spherical Waves

As is known, in the case of an individual travelling plane wave in a homogeneous boundless medium, pressure at any point in space is proportional to the particle velocity and is in phase with it. $p(t)$ and $V(t)$ are related by $V = \pm\, p/\rho c$. The '+' sign describes a wave travelling to the right (in the $+\,x$ direction), and the '–' sign a wave travelling in the $-\,x$ direction (where ρ is the density of the medium and c is the sound speed). This relationship also holds for pressure and particle velocity amplitudes in plane waves: $V_0 = p_0/\rho c$.

The individual travelling plane wave and the standing plane wave are fundamental wave objects that demonstrate possible connections between p and V. In the former, p and V are in phase; in the latter, p and V are π out of phase.

Let us consider the relationship between acoustic pressure and particle velocity in a spherical wave travelling in a homogeneous boundless medium. In a diverging monochromatic wave [1]:

$$p = \frac{a}{r} e^{-i(\omega t + kr)} = p_m e^{i(\omega t - kx - \varphi(r))}, \tag{A.9}$$

where $k = \omega/c$ is the wave number, $p_m = a/r$. Phase velocity c for spherical pressure waves is the same as phase velocity for plane waves. To find the particle velocity, we integrate the Euler equation:

$$V = -\int_{t_0}^{t} \frac{1}{\rho} \frac{\partial p}{\partial r} \mathrm{d}t = \frac{a}{r\rho c \cos\varphi} e^{i(\omega t - kx - \varphi(r))}, \tag{A.10}$$

where $\cos\varphi = \frac{kr}{\sqrt{1+k^2r^2}}$; $\sin\varphi = \frac{1}{\sqrt{1+k^2r^2}}$; $\tan\varphi = \frac{1}{kr}$.

In contrast to the plane wave, in the spherical wave (A.10) the particle velocity lags a phase angle $\varphi(r)$ behind the pressure, and the absolute value of the velocity amplitude is $|V_m| = \frac{|p_m|}{\rho c \cos\varphi}$, which is greater than $\frac{|p_m|}{\rho c}$. In the far field ($kr \gg 1$), $\cos\varphi \to 1$ and $\sin\varphi \to 0$, that is, ($\varphi \to 0$), and so the spherical wave in this case acquires the properties of a plane wave $V \to p/c$. In the far field, p and V decay as $1/r$. In the near field ($kr \ll 1$), $\cos\varphi \to kr$, $\sin \to 1$ and $\varphi \to \pi/2$. In this case, $p_m \to a/r$, but $V_m \to a/r^2$, and particle velocity lags $\pi/2$ behind pressure, as it does in a standing plane wave. Therefore, a diverging individual spherical wave changes its properties as a function of r. Amplitudes of p and V in the near field ($kr \ll 1$) are related by:

$$V(r) = \frac{p(r)e^{-i\varphi}}{\rho c \cos\varphi}.$$

This can be rearranged as

$$V(r) = \frac{p(r)}{\rho c \cos\varphi}(\cos\varphi - i\sin\varphi) = \frac{p(r)}{\rho c} - i\frac{p(r)}{\rho ckr} \tag{A.11}$$

The first term in (A.11) is the same as for a plane wave, meaning that p and V are in phase; the second term lags i/2 behind p. The first term is known as the active component of particle velocity V_a, and the second as the reactive component of particle velocity V_R. Our main conclusion is that in the near field of a spherical pulsating emitter, particle velocity has a phase shift relative to sound pressure, requiring the introduction of active $\boldsymbol{V}_a$ and reactive $\boldsymbol{V}_R$ components of particle velocity and, accordingly, active and reactive intensity. In the far field, a spherical field approximates the field of a plane travelling wave. Many researchers believed that measuring only acoustic pressure in the far field was sufficient to fully describe the acoustic field. However, as studies in mid- to late-twentieth century showed, a complete description of acoustic field in the shallow- and deepwater acoustic waveguide requires, as a minimum, simultaneous measurements of pressure and particle velocity at one point in space. In the general case, there is also a shift between $p(t)$ and $V(t)$ in the interference field of travelling plane waves, meaning that the reactive component of particle velocity is also non-zero. The phase shift causes the end of the particle velocity vector to trace an ellipse over the period $T = 2\pi/\omega$—in other words, within one period, the velocity vector has projections in all directions within the 2π angle. In a stationary field, $\boldsymbol{V}_{\mathrm{a}}$ and $\boldsymbol{V}_{\mathrm{R}}$ are constant vectors and can be written as:

$$\begin{aligned}\boldsymbol{V}_a &= \boldsymbol{i}\, V_{0,x} \cos \Delta\varphi_x + \boldsymbol{j}\, V_{0,y} \cos \Delta\varphi_y + \boldsymbol{k}\, V_{0,z} \cos \Delta\varphi_z \\ \boldsymbol{V}_R &= \boldsymbol{i}\, V_{0,x} \sin \Delta\varphi_x + \boldsymbol{j}\, V_{0,y} \sin \Delta\varphi_y + \boldsymbol{k}\, V_{0,z} \sin \Delta\varphi_z.\end{aligned} \tag{A.12}$$

Particle velocity vector $\boldsymbol{V}(t)$ rotates in the direction of the smaller angle between vectors $\boldsymbol{V}$a and $\boldsymbol{V}_{\mathrm{R}}$ in the plane of these vectors.

A.1.3 Analytic Signal

Although real physical processes are described by real-valued functions of time t, complex numbers are better suited to computations with experimental data, including the study of wavefields in acoustics. However, to replace the real-valued $p(t)$ or $\boldsymbol{V}(t)$ with their complex counterparts, such as $\tilde{p} = p_1(t) - ip_2(t)$, we need to establish an unambiguous relationship between $p_1(t)$ and $p_2(t)$. This connection is made by the Hilbert transforms. The complex function $\tilde{p}(t)$ constructed using the Hilbert transform from the real-valued function $p(t)$ is called its analytic signal. The Hilbert transform is used in the cross-processing of scalar and vector characteristics of the wavefield, especially in the study of their coherent properties. Mean values of $p_1(t)$ and $p_2(t)$ are also related by the Hilbert transform. It then follows that a stationary process must satisfy the condition $\langle p_1(t)\rangle_t = \langle p_2(t)\rangle_t = 0$. In other words, the processes must be stationary and the signals must have zero mean.

As is known, an arbitrary signal $s(t)$ with a known spectral density $S(\omega)$ can be uniquely written as a sum of two components each containing only positive or only negative frequencies:

$$\begin{aligned}s(t) &= \frac{1}{2\pi}\int_{-\infty}^{\infty} S(\omega)e^{i\omega t}\mathrm{d}\omega \\ &= \frac{1}{2\pi}\int_{-\infty}^{0} S(\omega)e^{i\omega t}\mathrm{d}\omega + \frac{1}{2\pi}\int_{-0}^{\infty} S(\omega)e^{i\omega t}\mathrm{d}\omega.\end{aligned} \tag{A.13}$$

Function

$$z_s(t) = \frac{1}{\pi}\int_{0}^{\infty} S(\omega)e^{i\omega t}\mathrm{d}\omega \tag{A.14}$$

describes real oscillations and is called the analytic signal. We will rearrange the first integral in (A.13) by substituting $\xi = -\omega$:

$$\frac{1}{2\pi}\int_{-\infty}^{0} S(\omega)e^{i\omega t}\mathrm{d}\omega = \frac{1}{2\pi}\int_{\infty}^{0} S(-\xi)e^{-i\xi t}$$

$$\mathrm{d}\xi = \frac{1}{2\pi}\int_{0}^{\infty} S(-\omega)e^{-i\xi t}\mathrm{d}\xi = \frac{1}{2}z_s^*(t), \tag{A.15}$$

where $z_s^*(t)$ is the complex conjugate of $z_s(t)$. This transforms (A.13) into

$$s(t) = \frac{1}{2}\left[z_s(t) + z_s^*(t)\right] = \mathrm{Re} z_s(t) \tag{A.16}$$

The imaginary part of the analytic signal

$$\mathrm{Im} z_s(t) = \frac{1}{2}\left[z_s(t) - z_s^*(t)\right] \tag{A.17}$$

is called the conjugate signal of the original signal $s(t)$.

Hence, the analytic signal

$$z_s(t) = s(t) + \mathrm{i}\hat{s}(t) \tag{A.18}$$

on the complex plane is a vector whose modulus and phase angle are functions of time.

A.1.4 Spectral Density of the Analytic Signal. Hilbert Transform

Let us investigate the spectral density of the analytic signal—that is, the function $Z_s(\omega)$ from which $z_s(t)$ can be found by the inverse Fourier transform:

$$z_s(t) = \frac{1}{2\pi}\int_{-\infty}^{\infty} Z_s(\omega)e^{i\omega t}\mathrm{d}\omega. \tag{A.19}$$

From (A.14) it follows that $Z_s(\omega)$ is non-zero only in the positive frequency region:

$$Z_s(\omega) = \begin{cases} 2S(\omega), & \omega > 0 \\ 0, & \omega < 0 \end{cases}. \tag{A.20}$$

If $\hat{S}(\omega)$ is the spectral density of the conjugate signal, then by virtue of linearity of the Fourier transform

$$Z_s(\omega) = S(\omega) + \mathrm{i}\hat{S}(\omega). \tag{A.21}$$

Therefore, (A.20) will be satisfied only if the spectral densities of the original and conjugate signals are related as follows:

$$\hat{S}(\omega) = -i\,\mathrm{sgn}(\omega)S(\omega) = \begin{cases} -iS(\omega), & \omega > 0 \\ iS(\omega), & \omega < 0 \end{cases}. \tag{A.22}$$

Equation (A.22) suggests that this transformation rotates the phases of all spectral components through – 90° for $\omega > 0$ and through + 90° for negative frequencies without changing their amplitudes. This transformation is known as the Hilbert transform. The symbolic notation is as follows:

$$\begin{array}{l} \text{forward transform } \hat{s}(t) = H[s(t)] \\ \text{inverse transform } s(t) = H^{-1}[\hat{s}(t)] \end{array}. \tag{A.23}$$

Consider a harmonic signal $s(t)$ in Fourier representation:

$$s(t) = \frac{1}{2\pi}\int_{-\infty}^{\infty} S(\omega)(\cos\omega t + i\sin\omega t)\mathrm{d}\omega. \tag{A.24}$$

The Hilbert transform of harmonic functions $\cos\omega t$ and $\sin\omega t$ is as follows:

$$\begin{aligned} H(\cos\omega t) &= \sin\omega t\,\mathrm{sgn}(\omega), \\ H(\sin\omega t) &= -\cos\omega t\,\mathrm{sgn}(\omega), \end{aligned} \tag{A.25}$$

where 'sgn' stands for the sign of ω.

A.1.5 Differential Vector Field Relations

The following familiar relations need to be quoted here to understand the theoretical text. Applying the differential vector nabla operator (del, or Hamilton operator)

$$\nabla = \boldsymbol{i}\frac{\partial}{\partial x} + \boldsymbol{j}\frac{\partial}{\partial y} + \boldsymbol{k}\frac{\partial}{\partial z}, \tag{A.26}$$

to scalar or vector variables of the field we get the following relations:

1. Gradient of the scalar acoustic pressure function

$$\nabla p = \boldsymbol{grad}\, p = \boldsymbol{i}\frac{\partial p}{\partial x} + \boldsymbol{j}\frac{\partial p}{\partial y} + \boldsymbol{k}\frac{\partial p}{\partial z} \tag{A.27}$$

$$\boldsymbol{grad}\, p \overset{\text{def}}{=} \nabla \cdot p.$$

2. Divergence of the vector $\boldsymbol{V}$

$$\nabla \cdot \boldsymbol{V} = \operatorname{div}\boldsymbol{V} = \frac{\partial V_x}{\partial x} + \frac{\partial V_y}{\partial y} + \frac{\partial V_z}{\partial z}$$

$$\operatorname{div}\boldsymbol{V} = \nabla \cdot \boldsymbol{V} \tag{A.28}$$

3. Rotation of the vector $\boldsymbol{V}$ (symbolically written as a vector product of the ∇ operator and $\boldsymbol{V}$)

$$\nabla \times \boldsymbol{V} = \boldsymbol{rot}\ \boldsymbol{V} = \boldsymbol{i}(\boldsymbol{rot}\ \boldsymbol{V})_x + \boldsymbol{j}(\boldsymbol{rot}\ \boldsymbol{V})_y + \boldsymbol{k}(\boldsymbol{rot}\ \boldsymbol{V})_z \tag{A.29}$$

$$\nabla \times \boldsymbol{V} \overset{\text{def}}{=} \boldsymbol{rot}\ \boldsymbol{V}$$

$$(\boldsymbol{rot}\ \boldsymbol{V})_x = \frac{\partial V_z}{\partial y} - \frac{\partial V_y}{\partial y},$$
$$(\boldsymbol{rot}\ \boldsymbol{V})_y = \frac{\partial V_x}{\partial z} - \frac{\partial V_z}{\partial x},$$
$$(\boldsymbol{rot}\ \boldsymbol{V})_z = \frac{\partial V_y}{\partial x} - \frac{\partial V_x}{\partial y}. \tag{A.30}$$

4. Scalar operator

$$\nabla^2 = \nabla \cdot \nabla = \Delta, \text{ where } \Delta \text{ is the Laplace operator}$$

$$\Delta = \frac{\partial^2}{\partial x^2} + \frac{\partial^2}{\partial y^2} + \frac{\partial^2}{\partial z^2} \overset{\text{def}}{=} \operatorname{div} \boldsymbol{grad}. \tag{A.31}$$

References

1. S.N. Rzhevkin. *A Course in the Theory of Sound* (in Russian) (MSU, Moscow, 1960)
2. I.N. Bronstein, K.A. Semendyaev. *Handbook of Mathematics* (in Russian) (Nauka, Moscow, 1981)

Appendix B
Fourth Statistical Moment of the Acoustic Vector Field

Building on the statistical theory of coherence developed in optics and radiophysics [1, 2], we will introduce a new relation in vector underwater acoustics: fourth statistical moment of the acoustic vector field. Correlation theory of coherence can be used to study correlation properties of the intensity vector (a second-order quantity), that is, a fourth-order statistical moment; this takes us beyond the second-order correlation theory. This approach greatly expands the possibilities in the study of vector acoustic fields.

The energy flux density vector (intensity vector) is a second-order statistical moment:

$$\boldsymbol{I}(f) = \langle p(f,t)\boldsymbol{V}(f,t)\rangle_t \tag{B.1}$$

where $p(f, t)$ and $\boldsymbol{V}(f, t)$ are instantaneous acoustic pressure and particle velocity vector, respectively; $<\ldots>_t$ means time averaging. Equation (B.1) is a cross-correlation function of two random processes: acoustic pressure $p(t)$ and particle velocity vector $\boldsymbol{V}(t)$, with a relative time shift $\tau = 0$. We assume that the acoustic field is stationary and ergodic, and $p(t)$ and $\boldsymbol{V}(t)$ are zero-mean values. We consider the signals to be monochromatic.

Switching to the complex plane, for a point of the field (where the receiver is located), we will write the time coherence function in Cartesian coordinates in the form:

$$\Gamma_{pV_i}(t) = \frac{\left\langle \tilde{p}(t)\tilde{V}_i^*(t)\right\rangle_t}{\sqrt{\left\langle \tilde{p}(t)\tilde{p}_i^*(t)\right\rangle\left\langle \tilde{V}_i(t)\tilde{V}_i^*(t)\right\rangle_t}} = \mathrm{Re}\,\Gamma_{pV_i}(t) + \mathrm{Im}\Gamma_{pV_i}(t). \tag{B.2}$$

V. A. Shchurov, *Movement of Acoustic Energy in the Ocean*,
https://doi.org/10.1007/978-981-19-1300-6

The argument of $\Gamma_{pV_i}(t)$ is $\theta(t) = \arctan\frac{\mathrm{Im}\,\Gamma_{pV_i}(t)}{\mathrm{Re}\,\Gamma_{pV_i}(t)}$, where $j^2 = -1$, $\tilde{p}(t)$, $\tilde{V}_i(t)$, $\tilde{V}_i^*(t)$ are analytic signals of acoustic pressure and components of particle velocity at a frequency ω, $i = x, y, z$. The coherence function has a range of $-1 \le \Gamma_{pV_i}(t) \le +1$. If $\mathrm{Re}\,\Gamma_{pV_i}(t) = \pm 1$, acoustic values $p(t)$ and $\boldsymbol{V}(t)$ are fully coherent. If $\mathrm{Re}\,\Gamma_{pV_i}(t) = 0$, $\mathrm{Im}\,\Gamma_{pV_i}(t) = \pm 1$, the field is incoherent (diffuse); and if $-1 < \mathrm{Re}\,\Gamma_{pV_i}(t) < +1$, $-1 < \mathrm{Im}\,\Gamma_{pV_i}(t) < +1$, the field is partially coherent. Quantities $p(t)$ and $\boldsymbol{V}(t)\{V_x(t), V_y(t), V_z(t)\}$ are random functions of time and coordinates.

Consider now the coherent properties of instantaneous intensity over time using higher-order as well as second-order variables in the form of (B.1) and (B.2).

We assume the observation time T to be much greater than the period T_0 of the carrier frequency ω_0 ($T > T_0 = 2\pi/\omega_0$). Let $\boldsymbol{I}(t)$ be the instantaneous intensity. Coherence $\Gamma_{pV_i}(t)$ is normalised intensity $I(t)$ in a frequency band and is a low-frequency function of time with the argument $\theta(t)$. Consider two combined receivers in acoustic field spaced $d > \lambda$ horizontally (Fig. B.1). The measurement process is synchronised across all receivers. Let $\boldsymbol{I}_1(t)$ be the measured intensity at point 1 and $\boldsymbol{I}_2(t)$ at point 2. The question arises, can the correlation of intensities $\boldsymbol{I}_1(t)$ and $\boldsymbol{I}_2(t)$ contain information about fluctuations of intensities at point 1 and point 2? We will write the correlation function of intensity in the form [2]:

$$\psi_I(\tau) = \langle I_1(t+\tau) I_2(t)\rangle - \langle I_1(t+\tau)\rangle\langle I_2(t)\rangle, \tag{B.3}$$

where τ is the relative time delay due to path difference Δl from the local source to each of the two receivers (Fig. B.1).

Therefore, correlation of intensities $\boldsymbol{I}_1(t)$ and $\boldsymbol{I}_2(t)$ requires the computation of a fourth-order statistical moment, which transcends the second-order correlation theory. The fourth statistical moment in (B.3) is the sum of pairwise products of the second moments [2]:

$$\begin{aligned}\left\langle p_1(t+\tau)V_1^*(t+\tau)p_2(t)V_2^*(t)\right\rangle &= \left\langle p_1(t+\tau)V_1^*(t+\tau)\right\rangle\left\langle p_2(t)V_2^*(t)\right\rangle \\ &+ \langle p_1(t+\tau)p_2(t)\rangle\left\langle V_1^*(t+\tau)V_2^*(t)\right\rangle \\ &+ \left\langle p_1(t+\tau)V_2^*(t)\right\rangle\left\langle V_1^*(t+\tau)p_2(t)\right\rangle \\ &= I_1 I_2 + |B_{12}(\tau)|^2 + \left|\tilde{B}_{12}(\tau)\right|^2,\end{aligned} \tag{B.4}$$

where $B_{12}(\tau) = \left\langle p_1(t+\tau)V_2^*(t)\right\rangle$ and $\tilde{B}_{12}(\tau) = \langle p_1(t+\tau)p_2(t)\rangle = 0$ are the first and the second correlation functions. The third term is zero because it describes looped energy flows (vortices of the intensity vector) [3].

As a result, (B.3) becomes $\psi_1(\tau) = |B_{12}(\tau)|^2 = I_1 I_2 |K_{12}(\tau)|^2$. But

$$K_{12}(\tau) = \frac{B_{12}(\tau)}{2\sqrt{I_1 I_2}} = |K_{12}(\tau)|e^{i\theta_{12}(\tau)}, \tag{B.5}$$

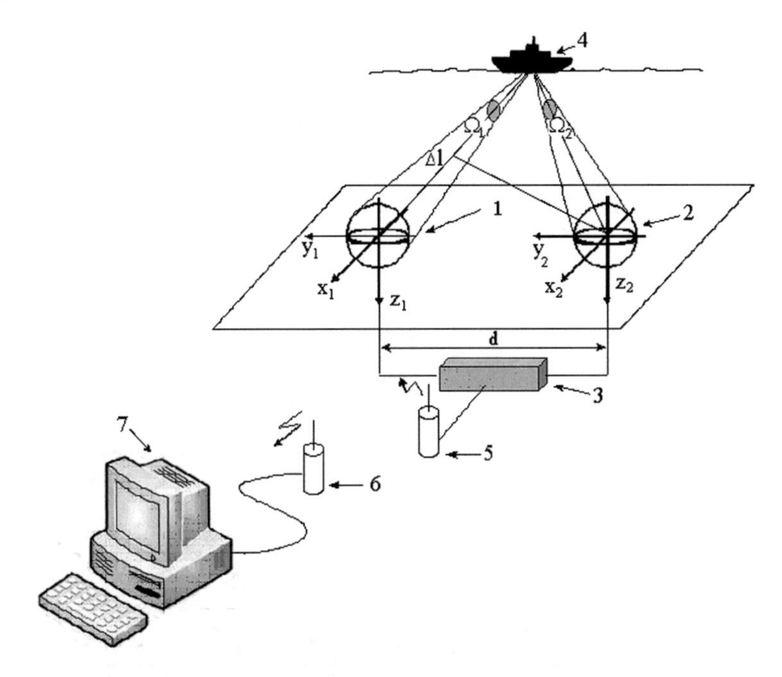

Fig. B.1 Experiment setup of the intensity interferometer (a counterpart of the Young–Rayleigh optical interferometer). Legend: 1, 2—combined receivers; 3—sealed container with electronics; 4—sound source, $\Delta l = \tau \cdot c$ is the path difference; $d = 300$ m is receiver spacing (interferometer base); 5, 6—transceivers; 7—multichannel digital data processing system

where $K_{12}(\tau)$ is the correlation coefficient of complex amplitudes. In its final form, the correlation function of intensity looks as follows:

$$\psi_I(\tau) = I_1 I_2 |K_{12}(\tau)| \cos\theta_{12}(\tau) \tag{B.6}$$

In (B.6), $I_1, I_2, K_{12}(\tau)$ are constants. Turning to Fig. B.1, it is evident that as the sound source moves relative to the first and second receivers of the interferometer, the path difference Δl, and thus the delay τ, will change, which will apply modulation of the form $\cos\theta_{12}(\tau)$ to the intensity correlation function $\psi_I(\tau)$.

Below are our experimental results from 2014. To test (B.6), we conducted a full-scale experiment in shallow water. The receiving system was an intensity interferometer [4] consisting of two vertical lines, each with two vertically spaced combined receivers (16 digital data channels). Local depth was ~ 30 m. The receivers were spaced 300 m apart horizontally. We investigated coherence of intensity measured by two combined receivers at depths of 15 m.

Frequencies ranged from 23 to 600 Hz. Figure B.2 is a correlogram of the y component of intensity in the case of a passing vessel. As evident from (B.6), the constant level signal is modulated by $\cos\theta_{12}(\tau)$. For comparison, Fig. B.3 plots $\mathrm{Re}\,\Gamma_{pV_y}(\tau)$ (B.2) for the y components of intensity of the first and second receivers.

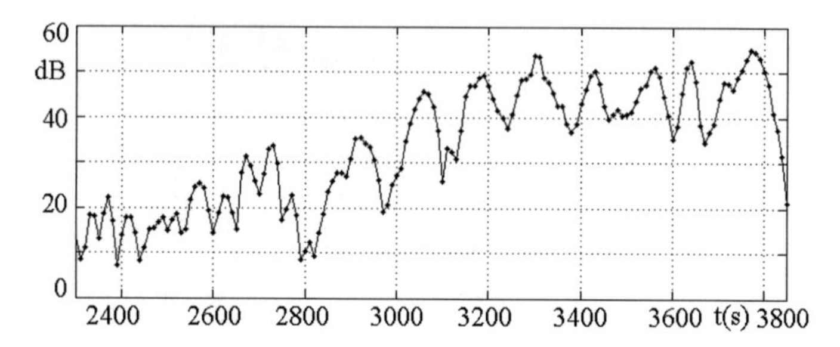

Fig. B.2 *Y* component of the intensity correlogram of the passing vessel. Frequency: 166 Hz, bandwidth: 6 Hz. Averaging time: 20 s. The decibel scale is arbitrary

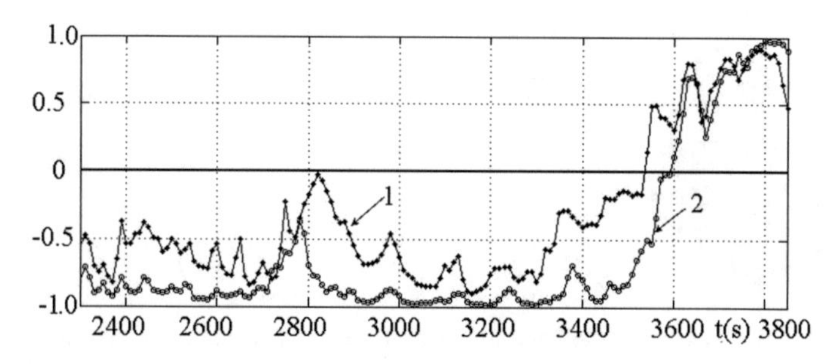

Fig. B.3 Time dependence of the y components of intensities $\mathrm{Re}I_{y_1}(t)$ and $\mathrm{Re}I_{y_2}(t)$ from the passing vessel at receiving points of the interferometer spaced 300 m apart. Frequency: 166 Hz, bandwidth: 6 Hz. Averaging time: 3 s. The '+' sign means that signal energy travels in the + *y* direction, and vice versa for '–'

A comparison of Figs. B.2 and B.3 shows that fluctuations of $\mathrm{Re}I_{y_1}(t)$ and $\mathrm{Re}I_{y_2}(t)$ transform into a common pattern of intensity correlation between points 1 and 2.

Therefore, statistical properties of $p(t)$ and $\boldsymbol{V}(t)$, via cross-correlation of $\mathrm{Re}\boldsymbol{I}_1(t)$ and $\mathrm{Re}\boldsymbol{I}_2(t)$, are reflected in the correlation properties of $\psi_I(\tau)$. Figure B.3 shows that field coherence at point 2 is higher than at point 1 for much of the time interval (2400–3800 s). The minus sign indicates that energy flows in the –*y* direction. Fluctuations of intensity (with dimensionless values converted to power decibels) do not exceed ± (3–5) dB. However,

Modulation of $\psi_I(\tau)$, on the other hand, reaches ± 25 dB. Therefore, $\psi_I(\tau)$ responds to changes in field coherence more than intensity does. Between 3400 and 3600 s, there is a change of sign from '–' to '+' of the *y* components of intensities (Fig. B.3), meaning that the sound source moves from the 1st to the 3rd coordinate quadrant of the combined receivers making up the interferometer. This transit doesn't affect the correlogram of $\psi_I(\tau)$, and neither should it, since the location of the coordinate system must not affect the coherent properties of acoustic field.

Investigation of complex acoustic processes and their degree of coherence using the fourth-order statistical moment of the intensity vector reveals entirely new, previously unknown information about the acoustic field of noise and signal, giving a new impetus to the theory of partial and complete coherence in vector acoustics. It is evident that this approach can find use in applications.

References

1. M. Born, E. Wolf. *Principles of Optics* (Russian edition) (Nauka, Moscow, 1957)
2. S.M. Rytov. *Introduction to Statistical Radiophysics* (in Russian) (Nauka, Moscow, 1976)
3. V.A. Shchurov, V.P. Kuleshov, A.V. Cherkasov. Vortex properties of the acoustic intensity vector in shallow water (in Russian), Akusticheskij Zhurnal (Acoustical Physics) **57**(6) (2011), pp. 837–843
4. V.A. Shchurov, S.G. Shcheglov, V.P. Kuleshov, E.N. Ivanov, E.S. Tkachenko. A hydroacoustic combined intensity interferometer (in Russian). 26th session of the Russian Acoustic Society combined with the 14th L.M. Brekhovskikh Ocean Acoustics Workshop (2013), pp. 335–338

Appendix C
Some International Publications (Patents and Articles) On Vector Acoustics in the Past Two Decades

Patent	Filed	Published	Applicant	Title
US5392258	Oct 12 1993	21 Feb 1995	The United States of America as represented by the Secretary of the Navy	Underwater acoustic intensity probe
US5930201A	27 Jan 1998	27 Jul 1999	The United States of America as represented by the Secretary of the Navy	Acoustic vector sensing sonar system
US6172940	27 Jan 1999	9 Jan 2001	The United States of America as represented by the Secretary of the Navy	Two geophone underwater acoustic intensity probe
US6370084B1	25 Jun 2001	9 Apr 2002	The United States of America as represented by the Secretary of the Navy	Acoustic vector sensor
US6697302B1	1 Apr 2003	24 Feb 2004	The United States of America as represented by the Secretary of the Navy	Highly directive underwater acoustic receiver
WO2005008193A2	9 Jul 2004	27 Jan 2005	Ken Deng	Acoustic vector sensor

(continued)

V. A. Shchurov, *Movement of Acoustic Energy in the Ocean*,
https://doi.org/10.1007/978-981-19-1300-6

(continued)

Patent	Filed	Published	Applicant	Title
US7066026	9 Jul 2004	27 Jun 2006	Wilcoxon Research, Inc.	Underwater acoustic vector sensor using transverse-response free, shear mode, PMN-PT crystal
WO2006137927A2	2 Nov 2005	28 Dec 2006	Gerald C. Lauchle	A rigidly mounted underwater acoustic inertial vector sensor
US7274622B1	23 May 2005	25 Sep 2007	The United States of America as represented by the Secretary of the Navy	Nonlinear techniques for pressure vector acoustic sensor array synthesis
CN100554896C	2 Feb 2005	28 Oct 2009	哈尔滨工程大学	High frequency small two-dimension coseismal column type vector hydrophone
US7536913B2	1 Nov 2005	26 May 2009	The Penn State Research Foundation	Rigidly mounted underwater acoustic inertial vector sensor
US20100316231A1	15 Jun 2009	Dec 16 2010	The Government of the US, as represented by the secretary of the Navy	System and method for determining vector acoustic intensity external to a spherical array of transducers and an acoustically reflective spherical surface
US20100265800A1		21 Oct 2010	Graham Paul Eatwell	Array shape estimation using directional sensors
US7839721B1	30 Jul 2008	23 Nov 2010	The United States of America as represented by the Secretary of the Navy	Modal beam processing of acoustic vector sensor data

(continued)

(continued)

Patent	Filed	Published	Applicant	Title
US8077540	15 Jun 2009	13 Dec 2011	The United States of America as represented by the Secretary of the Navy	System and method for determining vector acoustic intensity external to a spherical array of transducers and an acoustically reflective spherical surface
US8085622B2	31 Mar 2009	27 Dec 2011	The Trustees of the Stevens Institute of Technology	Ultra low frequency acoustic vector sensor
CN102353937B	6 Jul 2011	24 Apr 2013	哈尔滨工程大学	Single-vector active acoustic intensity averager
US8638956	29 Jul 2010	28 Jan 2014	Ken K. Deng	Acoustic velocity microphone using a buoyant object
US8873340B1	13 Feb 2013	28 Oct 2014	The United States of America as represented by the Secretary of the Navy	Highly directive array aperture
CN102914354B	Oct 26 2012	20 May 2015	哈尔滨工程大学	A three-dimensional combined hydrophone

Bibliography

1. M.J. Berliner, Acoustic particle velocity sensors: design, performance, and applications. J. Acoust. Soc. Am. **100**(6), 3478–3479 (1966)
2. T.B. Gabrielson, et al., A simple neutrally buoyant sensor for direct measurement of particle velocity and intensity in water. J. Acoust. Soc. Am. **97**(4), 2227–2237 (1995)
3. A. Nehorai, E. Paldi, Acoustic vector-sensor array processing. IEEE Trans. Sig. Proc. **42**(9), 2481–2491 (1994)
4. B.A. Cray, A.H. Nuttall, Directivity factors for linear arrays of velocity sensors. J. Acoust. Soc. Am. **110**(1), 324–331 (2001)
5. G.L. D'Spain, W.S. Hodgkiss, G.L. Edmonds, Energetics of the deep ocean's infrasonic sound field. J. Acoust. Soc. Am. **89**(3), 1134–1158 (1991)
6. H. Teutsch, W. Kellermann, Acoustic source detection and localization based on wavefield decomposition using circular microphone arrays. J. Acoust. Soc. Am. **120**(5), 2724–2736 (2006)
7. H. Teutsch, W. Kellermann, EB-ESPRIT: 2D localization of multiple wideband acoustic sources using eigen-beams. ICASSP, III-89–III-92 (2005)
8. H. Teutch, W. Kellermann, Eb-Espirit: 2D localization of multiple wideband acoustic sources using eigen-beams, in *2005 International Conference on Acoustics, Speech, and Signal Processing (ICASSP 2005)*, vol. 3 (IEEE, March 18–23, 2005, Philadelphia, Pennsylvania), pp. III-89–III-92
9. H. Teutsch, W. Kellermann, Acoustic source detection and localization based on wavefield decomposition using circular microphone arrays. J. Acoust. Soc. A. **120**(5), 2724–2736 (2006)
10. J.A. Clark, G. Tarasek, *Localization of Radiating Sources along the Hull of a Submarine Using a Vector Sensor Array. Oceans'06* (IEEE, Boston, MA, Sep 18–21, 2006)
11. J. Meyer, G. Elko, A highly scalable microphone array based on an orthonormal decomposition of the soundfield. ICASSP, II-1781–II-1784 (2002)
12. J.A. Clark, D. Huang, High resolution angular measurements with single vector sensors and arrays. J. Acoust. Soc. Am. **123**(5), 3006 (2008)
13. J.A. Clark, G. Tarasek, Localization with vector sensors in inhomogeneous media. 153 rd meeting of the acoustical society of America. J. Acoust. Soc. **121**(5), 3070 (2007)
14. J.A. Clark, G. Tarasek, Radiated noise measurements with vector sensor arrays. 151st meeting of the acoustical society of America. J. Acoust. Soc. Am. **119**(5), 3444 (2006)
15. J.A. Clark, Calibration of vector sensors. J. Acoust. Soc. Am. **123**(5), 3347 (2008)
16. J.A. Clark, G. Tarasek. Localization of *Radiating Sources along the Hull of a Submarine Using a Vector Sensor Array. Oceans'06* (IEEE, Boston, Massachusetts, Sep 18–21, 2006), p. 3

V. A. Shchurov, *Movement of Acoustic Energy in the Ocean*,
https://doi.org/10.1007/978-981-19-1300-6

17. J.A. Clark, Rapid communication high-order angular response beamformer for vector sensors. J. Sound Vibr. **318**(3), 417–422 (2008)
18. K.B. Smith, A.V. van Leijen, Steering vector sensor array elements with linear cardioids and non linear hippioids. J. Acoust. Soc. Am. **122**(1), 370–377 (2007)
19. K.B. Smith, A. Vincent van Leijen, Steering vector sensor array elements with linear cardioids and non-linear hippioids. J. Acoust. Soc. Am. **122**(1), 370–377 (2007)
20. M. Hawkes, A. Nehorai, Acoustic vector-sensor beamforming and capon direction estimation. IEEE Trans. Sig. Proc. **46**(9), 2291–2304 (1998)
21. M.J. Berliner, J.F. Lindberg, *Acoustic Particle Velocity Sensors: Design, Performance and Applications* (Woodbury, N.Y., 1996)
22. N. Qi, T. Tian, Acoustic vector hydrophone array supergain energy flux beamforming, in *Eighth International Conference on Signal Processing (ICSP'06)*, Nov 16–20, 2006, Guillin, China, p. 4
23. R. Hickling, W. Wei, R. Raspet, Finding the direction of a sound source using a vector sound-intensity probe. J. Acoust. Soc. Am. **94**(4), 2408–2412 (1993)
24. U.S. Appl. N. 61/070,617, filing date Mar. 13, 2008, invention title "Modal Beam Processing of Acoustic Vector Sensor Data," sole inventor Joseph A. Clark
25. V.A. Shchurov, A.V. Shchurov, Noise immunity of a combined hydroacoustic receiver. Acoust. Phys. **48**(1), 98–106 (2002)
26. G.L. D'Spain, Relationship of underwater acoustic Intensity measurement to beamforming. Can. Acoust. **22**(3), 157–158 (1994)
27. Hawkes et al., Bearing estimation with acoustic vector sensor arrays. AIP Conf. Proc. **1996**(368), 345–358 (1996). ISSN 0094 243X
28. P.C. Herdic, B.H. Houston, M.H. Marcus, E.G. Williams, A.M. Baz, The vibro-acoustic response and analysis of a full-scale aircraft fuselage section for interior noise reduction. J. Acoust. Soc. Am. **11**(6), 3667–3678 (2005)
29. B.H. Houston, M.H. Marcus, J.A. Bucaro, E.G. Williams, D.M. Photiadis, The structural acoustics and active control of interior noise in a ribbed cylindrical shell. J. Acoust. Soc. Am. **99**(6), 3497–3512 (1996)
30. B.J. Sklanka, J.R. Tuss, R.D. Buehrle, J. Klos, E.G. Williams, N. Valdivia, Acoustic source localization in aircraft interiors using microphone array technologies. Paper No. AIAA-2006-2714, in *12th AIAA/CEAS Aeroacoustics Conference*, Cambridge MA, May 2006
31. N.P. Valdivia, E.G. Williams, Reconstruction of the acoustic field using patch surface measurements, in *Proceedings Thirteenth International Congress on Sound and Vibration*, Vienna, Austria, Jul 2006
32. E.G. Williams, Continuation of acoustic near-fields. J. Acoust. Soc. Am. **113**(3), 1273–1281 (2003)
33. E.G. Williams, Regularization methods for near-field acoustical holography. J. Acoust. Soc. Am. **110**(4), 1976–1988 (2001)
34. E.G. Williams, et al., *Tracking Energy Flow Using a Volumetric Acoustic Intensity Imager (VAIM). Presentation Slides, Inter-Noise 2006*, Hawaii, Dec 5, 2006, pp. 1–25
35. E.G. Williams, *Fourier Acoustics: Sound Radiation and Nearfield Acoustical Holography* (Academic Press Inc., 1999), Chap. 7, pp. 235–249
36. E.G. Williams, B.H. Houston, P.C. Herdic, Fast Fourier transform and singular value decomposition formulations for patch near-field acoustical holography. J. Acoust. Soc. Am. **114**(3), pp. 1322–1333
37. E.G. Williams, B.H. Houston, P.C. Herdic, S.T. Raveendra, B. Gardner, Interior near-field acoustical holography in flight. J. Acoust. Soc. Am. **108**(4), 1451–1463 (2000)
38. E.G. Williams, N. Valdivia, P.C. Herdic, J. Klos, Volumetric acoustic vector intensity imager. J. Acoust. Soc. Am. **120**(4), 1887–1897 (2006)
39. E.G. Williams, N.P. Valdivia, J. Klos, *Tracking Energy Flow Using a Volumetric Acoustic Intensity Imager (VAIM).* Paper presented at Inter-Noise 2006, Hawaii, Dec 3–6, 2006, pp. 1–10
40. E.G. Williams, Volumetric Acoustic Intensity Probe. 2006 NRL Review, Dec 2006, pp. 110–111

41. E.G. Willliams, A volumetric acoustic intensity probe based on spherical nearfield acoustical holography, in *Proceedings Thirteenth International Congress on Sound and Vibration*, Vienna, Austria, Jul 2006
42. H. Cox, A.B. Baggeroer, Performance of vector sensors in noise. J. Acoust. Soc. Am. **114**, 24–26 (2003)
43. C.R. Greene, M.W. McLennan, R.G. Norman, T.L. McDonald, R.S. Jakubczak, W.J. Richardson, Directional frequency and recording_DIFAR_sensors in seafloor recorders to locate calling bowhead whales during their fall migration. J. Acoust. Soc. Am. **116**, 799–813 (2004)
44. M.A. McDonald, DIFAR hydrophone usage in whale research. Can. Acoust. 155–160 (2004)
45. C.M. Traweek, Optimal Spatial Filtering for Design of a Conformal Velocity Sonar Array. Ph.D. thesis, The Pennsylvania State University, 2003
46. C.M. Traweek, A collaborative roadmap for vector sensor towed arrays enabled by piezocrystal materials. ONR powerpoint presentation, Office of Naval Research, 2004
47. Wilcoxon Research "The vector sensor," specification sheet, www.wilcoxon.com (2004)
48. J. Meyer, G. Elko, A highly scalable spherical microphone array based on an orthonormal decomposition of the soundfield, in *Proceedings of IEEE ICASSP*, vol. 2, Orlando, FL, USA, May 2002, pp. 1781–1784
49. B. Rafaely, Anaysis and design of spherical microphone arrays. IEEE Trans. Speech Audio Process. **13**(1), 135–143 (2005)
50. Z. Li, R. Duraiswami, Flexible and optimal design of spherical microphone arrays for beamforming. IEEE Trans. Speech Audio Process. **15**(2), 702–714 (2007)
51. H. Sun, S. Yan, U.P. Svensson, Robust minimum sidelobe beamforming for spherical microphone arrays. IEEE Trans. Speech Audio Process. **19**(4), 1045–1051 (2011)
52. S. Yan, H. Sun, P. Svensson, X. Ma, J.M. Hovem, Optimal modal beamforming for spherical microphone arrays. IEEE Trans. Speech Audio Process. **19**(2), 361–371 (2011)
53. H. Teutsch, W. Kellermann, Acoustic source detection and localization based on wavefield decomposition using circular microphone arrays. J. Acoust. Soc. Am. **120**(5), 2724–2736 (2006)
54. A.M. Torres, M. Cobos, B. Pueo, J.J. Lopez, Robust acoustic source localization based on modal beamforming and timecfrequency processing using circular microphone arrays. J. Acoust. Soc. Am. **132**(3), 1511–1520 (2012)
55. T.D. Abhayapala, A. Gupta, Higher order differential-integral microphone arrays. J. Acoust. Soc. Am. **127**(5), EL227–EL233 (2010)
56. S. Doclo, M. Moonen, Superdirective beamforming robust against microphone mismatch. IEEE Trans. Speech Audio Process. **15**(2), 617–631 (2007)
57. E. De Sena, H. Hacihabiboglu, Z. Cvetkovic, On the design and implementation of higher order differential microphones. IEEE Trans. Speech Audio Process. **20**(1), 162–174 (2012)
58. J. Chen, J. Benesty, C. Pan, On the design and implementation of linear differential microphone arrays. J. Acoust. Soc. Am. **136**(6), 3097–3113 (2014)
59. Y. Yang, C. Sun, C. Wan, Theoretical and experimental studies on broadband constant beamwidth beamforming for circular arrays, in *Proceedings of OCEANS*, San Diego, CA, Sep 2003, pp. 1647–1653
60. Y. Ma, Y. Yang, Z. He, K. Yang, C. Sun, Y. Wang, Theoretical and practical solutions for high-order superdirectivity of circular sensor arrays. IEEE Trans. Ind. Electron. **60**(1), pp. 203–209 (2013)
61. Y. Wang, Y. Yang, Y. Ma, Z. He, Robust high-order superdirectivity of circular sensor arrays. J. Acoust. Soc. Am. **136**(4), 1712–1724 (2014)
62. A. Trucco, F. Traverso, M. Crocco, Broadband performance of superdirective delay-and-sum beamformers steered to end-fire. J. Acoust. Soc. Am. **135**(6), EL331–EL337 (2014)
63. B.A. Cray, V.M. Evora, A.H. Nuttall, Highly directional acoustic receivers. J. Acoust. Soc. Am. **113**(3), 1526–1532 (2003)
64. D.J. Schmidlin, Directionality of generalized acoustic sensors of arbitrary order. J. Acoust. Soc. Am. **121**(6), 3569–3578 (2007)

65. J.A. McConnell, S.C. Jensen, J.P. Rudzinsky, Forming first and second-order cardioids with multimode hydrophones, in *Proceedings of OCEANS*, Boston, MA, Sep 2006
66. J.C. Shipps, B.M. Abraham, The use of vector sensors for underwater port and waterway security, in *Proceedings of Sensors Industry Conference*, New Orleans, LA, Jan 2004, pp. 41–44
67. M.T. Silvia, R.T. Richards, A theoretical and experimental investigation of low-frequency acoustic vector sensors, in *Proceedings of OCEANS*, vol. 3, Biloxi, MS, 2002, pp. 1886–1897
68. J.F. McEachern, J.A. McConnell, J. Jamieson, D. Trivett, ARAP deep ocean vector sensor research array, in *Proceedings of OCEANS*, Boston, MA, Sep 2006
69. G.L. D'Spain, J.C. Luby, G.R. Wilson, R.A. Gramann, Vector sensors and vector sensor line arrays: comments on optimal array gain and detection. J. Acoust. Soc. Am. **120**(1), 171–185 (2006)
70. N. Zou, A. Nehorai, Circular acoustic vector-sensor array for mode beamforming. IEEE Trans. Signal Process. **57**(8), 3041–3052 (2009)
71. B. Gur, Particle velocity gradient based acoustic mode beamforming for short linear vector sensor arrays. J. Acoust. Soc. Am. **135**(6), 3463–3473 (2014)
72. H.L. Van Trees, *Optimum Array Processing: Part IV of Detection, Estimation, and Modulation Theory* (Wiley, 2002), Chap. 4, pp. 231–322
73. L.E. Kinsler, A.R. Frey, A.B. Coppens, J.V. Sanders, *Fundamentals of Acoustics*, 4th edn. (Wiley, 2000), Chap. 4, pp. 113–148
74. X. Guo, S. Yang, S. Miron, Low-frequency beamforming for a miniaturized aperture three-by-three uniform rectangular array of acoustic vector sensors. J. Acoust. Soc. Am. **138**, 3873–3883 (2015)

Printed in the United States
by Baker & Taylor Publisher Services